The Elements of
FIBER OPTICS

S. L. Wymer Meardon

Regents/Prentice Hall
Englewood Cliffs, New Jersey 07632

Library of Congress Cataloging-in-Publication Data

Meardon, S. L. Wymer (Susan L. Wymer), 1956–
   The elements of fiber optics / by S.L. Wymer Meardon.
     p.   cm.
   Includes bibliographical references and index.
   ISBN 0-13-249699-2
   1. Optical communications. 2. Fiber optics. 3. Optical fibers.
 I. Title.
   TK5103.59.M43   1993
   621.36'92—dc20
                            91-48012
                             CIP

Production editor: *Adele M. Kupchik*
Acquisitions editor: *Holly Hodder*
Cover designer: *Ben Santora*
Prepress buyer: *Ilene Levy*
Manufacturing buyer: *Ed O'Dougherty*
Editorial assistants: *Cathy Frank and Elizabeth O'Brien*

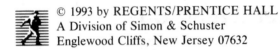

© 1993 by REGENTS/PRENTICE HALL
A Division of Simon & Schuster
Englewood Cliffs, New Jersey 07632

Printed in the United States of America

10 9 8 7 6 5 4 3 2 1

ISBN 0-13-249699-2

Prentice-Hall International (UK) Limited, *London*
Prentice-Hall of Australia Pty. Limited, *Sydney*
Prentice-Hall Canada Inc., *Toronto*
Prentice-Hall Hispanoamericana, S.A., *Mexico*
Prentice-Hall of India Private Limited, *New Delhi*
Prentice-Hall of Japan, Inc., *Tokyo*
Simon & Schuster Asia Pte. Ltd., *Singapore*
Editora Prentice-Hall do Brasil, Ltda., *Rio de Janeiro*

This book is dedicated to the memory of my father.
He always knew I would be an engineer.

Contents

Preface

The Elements of Fiber Optics has been written to provide a more practical exposure to fiber optics. The book assumes a knowledge of ac and dc circuits as well as transistor biasing and the associated circuitry. This book is intended for a course in the second or third year of an electronic engineering technology degree program. The book can also be used in the electrical installation industry as an introduction to fiber optics.

The book is arranged as follows:

Chapter 1	gives a short history of fiber optic cable technology. Fiber optic cable and electrical cable are also compared.
Chapter 2	presents the basic principles of light using Snell's law and its relationships to fiber optic cable.
Chapter 3	introduces the reader to the various types of fiber and some of the parameters associated with fiber specifications.
Chapter 4	presents the fiber manufacturing process, cable design, and cable testing.
Chapter 5	introduces connectors, splices, and couplers.
Chapter 6	introduces and compares optical sources for fiber optic transmission.
Chapter 7	introduces different photodetectors and the parameters associated with them.
Chapter 8	presents different transmission schemes for fiber optic systems.

Chapter 9	introduces practical design of optical sources and photodetectors for use in fiber optic systems.
Chapter 10	presents system architecture and different types of networks, such as local area networks, metropolitan area networks, and wide area networks.
Chapter 11	provides insight into designing complete fiber optic systems by the use of analytical worksheets.
Chapter 12	introduces various installations, techniques, and testing of complete fiber optic systems.
Appendix A	provides commonly used constants and a discussion of the use of decibels.
Appendix B	provides a short list of the standards for fiber optics.

S.L. Wymer Meardon

Acknowledgments

A book never gets written without the help, love, and understanding of everyone around you. The family and editor have to endure the excuses, comments, and surliness that somehow come with writing a book. My book is no different: A lot of people had to have the patience to endure this sojourn of mine and I want to thank them.

To Douglas and Patricia, my son and daughter: Thanks for the time and patience you gave me to finish this. You always make me realize that there are more things in life than work.

To Dan, my husband: Thanks for the help and encouragement.

To my mother: Thanks for the support, encouragement, and understanding that have endured throughout my life.

To Sharon Jacobus: Thanks for the encouragement and "Just do it" attitude.

To Elizabeth O'Brien: Thanks for keeping the book rolling and your patience with my never-ending questions.

To Cliff Dobbins: A good student who scrutinized my book and a nice guy.

Finally, to my students: I started writing this book because I could not find a good one to use. I hope my efforts have been worthwhile.

S.L. Wymer Meardon

1

Introduction to Fiber Optics

CHAPTER OBJECTIVES

The student will be able to:

- Describe significant events that lead to a functioning fiber optic cable.
- Describe the advantages and disadvantages of using fiber optics.
- Discuss applications in the telephone industry, the military, television, computers, automotive, and the wired city.

1.1 SHORT HISTORY OF FIBER OPTICS

In the next decades, the world will change due to fiber optic technology. The impact of fiber optics has already been felt in telephone communications, but the average homeowner could see an explosion of offerings, from home grocery shopping to current movies played in the home. The reason for such a change will be because of a strand of glass that is smaller than a human hair. The glass can transmit signals over long distances without degrading the signal as much as copper. However, that is for the future. A look at the past will show that in a few short years, fiber optics made great strides in evolving from an idea to a real transmission medium.

John Tyndall, in 1854, demonstrated that light could follow a curved path. In a presentation before the Royal Academy in London, Tyndall had water flowing out of the bottom of one container into another container below it. This flow of water made a parabolic path through the air. When a beam of light was aimed at the top container, the light illuminated the container and followed a zigzag path inside the curved walls of the water. This was an important step toward having light in a guided path.

Alexander Graham Bell, in 1880, took the idea further and studied the photophone, a device that could theoretically transmit speech on a beam of light. The voice vibrated a mirror that was directed to a selenium parabolic reflector some 200 meters (m) away. The selenium reflector detected the variations in the intensity of the light from the mirror. This intensity was transformed into a current that ran a speaker. Figure 1-1 shows a schematic of the photophone.

Tyndall proved that light could be guided, and Bell confirmed that speech could be transmitted on a beam of light. There were problems with Tyndall's demonstration. Because water created a rough boundary, light would leak out. Bell's photophone could travel only a short distance. If the light could be confined in some manner, the boundary and distance problem could be solved. This led to the idea of a two-layer glass rod (Figure 1-2). Light could be carried in the inner layer (the core) and prevented from leaking out by the outer layer (the cladding). In the 1950s, the first flexible fiberscope was invented by Harry Hopkins and Narinder Kapany. Kapany, who went on to invent the first practical glass-coated fiber, coined the term *fiber optics*.

In 1964, Charles K. Kao and George A. Hockman, researchers at the Standard Telecommunications Laboratories in England, "saw the light" so to speak. This was the first time that light was transmitted down an optical waveguide. The loss is measured in decibels per kilometer (dB/km). See Appendix A for a discussion of decibels. The loss is the difference in the

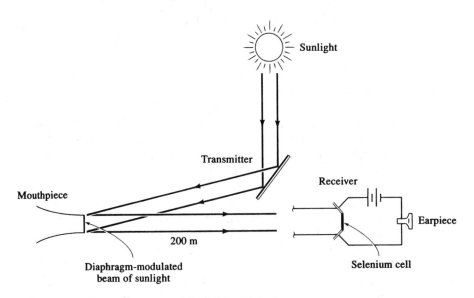

Figure 1-1 Photophone.

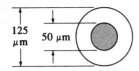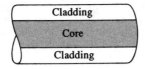

Figure 1-2 Basic fiber structure.

amount of light that went in versus the light that came out. At the time, the losses amounted to 1000 (dB/km), which by today's standards are astronomical. Work had to be done to get those losses down. Three years later at Corning Glass Works in New York, Robert Maurer, Donald Keck, and Peter Schultz produced the first fiber with losses of less than 20 dB/km. Today, losses typically are from 0.2 to 2.0 dB/km, depending on the type of cable and the wavelength of operation.

1.2 PROS AND CONS

Why did these people invest so much time, energy, and money in developing the fiber optic cable? They did primarily because the cable has some unique properties associated with it: large bandwidth, small size, freedom from electromagnetic pulses, signal confinement, and low cost. Table 1-1 summarizes the advantages of using fiber optic cable compared to copper cable.

1.2.1 Advantages

One of the greatest advantages of fiber optic cable over microwave systems and conventional wiring (twisted pair, coax, and copper cable) is the *information-carrying capacity* of fiber optics. Information-carrying capacity is normally expressed in terms of the amount of data that can be transferred through the cable in a given period of time, or information bandwidth. The higher the bandwidth, the greater the information-carrying capacity. Potential band-

TABLE 1-1 Advantages of Using Fiber Optic Cable

Characteristic	Advantage
Greater bandwidth	Can be used at high data rates
Smaller size	Can be installed in existing conduit
Signal confinement	No crosstalk or hum; secure communications
Freedom from electromagnetic pulses	No interference from lightning or ground loops
Low cost	Less expensive per installed voice circuit
Security	Hard to tap without detection

width is a function of the transmission frequency. For conventional wiring, the bandwidth is much more limited because of signal attenuation and other problems that will be discussed in more detail.

Example 1-1

A television broadcast data rate can be calculated by adding the sums of the analog, video, and sound track data to get the total bandwidth. The analog bandwidth is 3.5 megahertz (MHz). This must be sampled at twice the highest frequency contained in the signal, called the *Nyquist rate*, therefore 7.0 MHz. The signal is encoded in 9 bits per second (bps). The sound bandwidth is 15.75 kilohertz (kHz) sampled at the Nyquist rate and is encoded in 8 bps. What is the total bandwidth?

SOLUTION

$$BW_{video} = (3.5 \text{ MHz})(2)(9 \text{ bps}) = 63 \text{ Mbps}$$
$$BW_{sound} = (15.75 \text{ kHz})(2)(8 \text{ bps}) = 252 \text{ kbps}$$
$$\text{Total BW} = 63 \text{ Mbps} + 252 \text{ kbps} = 63,252,000 \text{ bps}$$
$$= 63.30 \text{ Mbps}$$

This signal could be put on one fiber optic cable, although the fiber could potentially carry gigabits per second.

One common digital application where fiber optics can be compared to conventional wiring relates to the telephone industry's T-carrier system. A T1 carrier has an information-carrying capacity of 1.544×10^6 bps or 24 voice channels. Subsequently higher T-carrier rates are multiples of the T1 level as shown in Table 1-2. The higher T1 rates, called *multiplexed signals*, send more than one signal or voice channel over the same wire. It is important to telephone companies to conveniently transfer as many voice channels as possible from one point to another.

TABLE 1-2 Specifications for the T-Carrier Telephony System

System	Rate (Mbps)	Multiplexed	Voice Channels
T1	1.544	T1	24
T1C	3.152	2(T1)	48
T2	6.312	4(T1)	96
T3	44.736	7(T2), 28(T1)	672
T3C	91.053	2(T3), 56(T1)	1344
T4	274.176	6(T3), 168(T1)	4032
T5	560.160	2(T4), 336(T1)	8064

With conventional wiring systems, the voice channels transmitted in a T-carrier would have to be split among many individual pairs of wire, but with fiber optics many voice channels can be transmitted over one fiber.

Example 1-2

A standard telephone cable has 900 twisted pairs and is 70 millimeters (mm) in diameter. Each pair carries 24 (T1 rate) voice channels. A fiber optic cable contains 144 fibers and has a total diameter of 12.7 mm. Each fiber can carry 672 (T3 rate) voice channels. Calculate the total number of calls for the twisted-pair wire and the fiber optic cable.

SOLUTION

For the twisted pair,

$$\text{number of calls} = \text{number of wires} \times \text{number of voice channels}$$
$$= 900 \times 24$$
$$= 21,600 \text{ calls}$$

For the fiber optic cable,

$$\text{number of calls} = \text{number of fibers} \times \text{number of voice channels}$$
$$= 144 \times 672$$
$$= 96,768 \text{ calls}$$

The fiber optic cable is a lot smaller in size than conventional wire cables. The fiber itself is about the size of a human hair, which is roughly 100×10^{-6} m. Add to that some type of protective coating, and the fiber is still smaller than wire. For example, a fiber optic cable that is roughly 0.25 inch (in.) in diameter will carry the same amount of information as a 3-in. bundle of 900 pairs of copper cable. In metropolitan areas, most telephone cabling is put through underground ducts, which are often already crowded or full. If additional telephone lines are needed, the phone company cannot use conventional wire because of its size. Adding new capacity, which is based on fiber optics technology, is often the only choice for the telephone companies.

Another problem with placing so many copper cables together is called *crosstalk*, defined as the interference of signals between adjacent wire pairs. Fiber optic cable is not a conductive medium, nor are the signals on fiber optic cable susceptible to external electric signals. Therefore, it does not give off any electrical signal that might interfere with the cable next to it. This is a useful characteristic because it eliminates the added cost of shielding the cable from other signals that could affect data transmission, especially at high data rates. Fiber optics has also been used for communications near high-voltage lines for the same reason. The electromagnetic fields present around high-voltage lines do not affect the data being transferred in a fiber optic cable. If a conventional wiring system were to be used, it would often be impossible to send or receive data near high-voltage lines.

Because fiber optic cable is not an electrically conductive medium, there are few problems with ground loops. That is, the fiber optic transmitter and receiver do not have to have a common ground between them. The cable terminations are isolated from each other. With conventional wiring, electrical isolation is not easily achieved. Also, since the signal is confined to the fiber, external pulses do not affect the fiber. Therefore, lightning, some alpha radiation, and some gamma radiation do not affect the data being transferred. The fiber cable is also immune from nuclear explosions and electromagnetic pulses (EMPs). In a U.S. Navy study of the nuclear radiation susceptibility of fiber optic cable, the cable was deemed to survive a radiation explosion, although there was some coloration change to brown. The attenuation of the cable was larger, but the electronics failed before the fiber. The military prefers to use fiber optic cable for secure communications. Although tapping is possible, detection of a monitored line cannot go unnoticed because of the signal loss in the cable.

There is one more advantage to fiber optics of which most people are not aware. The fiber optic cable that has been selected for a system is usually capable of carrying a higher bandwidth than that required by the signal that is being transmitted. This means that when the needs of the communication link change, such as when a higher bandwidth is needed, the fiber cable will not have to be replaced. Only the transmitter and the receiver will be changed. Not having to replace the fiber could lead to greater savings.

1.2.2 Disadvantages

The disadvantages of using fiber optic cables are few but must be mentioned. Some types of cable are harder to join, especially if using splicing. Splicing can be more time consuming to accomplish, thereby increasing the cost of installation of the cable.

Since fiber optic cable is nonconductive, if an electrical communications link is needed at a remote site, an additional conducting member has to be added to the total cable configuration. The additional conducting member for an underwater application is a copper wire used to send a 1.5-ampere (A) current at 2500 volts (V) to repeaters. The *repeater* is a combination receiver, amplifier, and transmitter used in longer transmissions (over 30 nautical miles) to regenerate the signal. Water intrusion can cause the fiber to decay over a long time period. Water breaks down the glass, making the glass brittle and eventually disintegrating.

Lack of standards and knowledge are also a problem. Standards for implementation of networks, signal levels, and more are still in committees. Certain standards for connectors, test equipment, and measurements have seemed to be set, but as always there will be new ways and ideas to change the de facto standards. Lack of knowledge of the industry and installation techniques has been a major issue. Many fiber manufacturing companies have created classes and seminars on the subject. All of the disadvantages can be solved or overcome but must not be overlooked.

1.3 BASIC FIBER OPTIC LINK

A fiber optic system link converts the electrical signal into light, transmits the light through the fiber, and converts the light back into an electrical signal. The input electrical signal can be either an analog (sound, voice, video camera) or digital format (on or off). A typical link, as depicted in Figure 1-3, has five major parts that send communications from one point to another point: encoder, light source or transmitter, the optical fiber, a light detector or receiver, and the decoder. The driver converts the signal, whether analog (varies the intensity of the light) or digital (turns the source on or off), to a format that is useful to the light source or transmitter. The light source takes the electrical signal from the driver and converts it to a light pulse. Since the light source is directly coupled to the optical fiber, the light pulses will be guided down the length of the fiber optic cable to the coupled light detector. The light pulses strike the surface of the light detector and are converted to an electrical current corresponding to the intensity of the light that was transmitted. The signal now is decoded and is a replica of the original pulse.

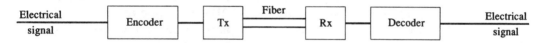

Figure 1-3   Basic fiber optic link.

1.4 APPLICATIONS

Fiber optics has become the method of choice for telecommunications because of the advantages and ease in which a system can be used. The telephone, automobile, military, television, and computer industry have benefited the most from the advances made in the technology. The telephone industry is the most prominent user of fiber optics. Major systems have been installed in Chicago, Orlando (Disney World), and New York to handle the telephone traffic in those areas. The most significant system will be installed by the late 1990s. This 611 mile link will connect Washington, D.C., Philadelphia, New York, Boston, Chicago, and Miami. The link will carry 80,000 simultaneous calls over a 0.5 in.-diameter fiber optic cable.

Not only are the telephone companies linking cities but they are linking continents. In 1988, the first fiber optic cable, the TAT-8 or Transatlantic Telephone, linked the United States, Great Britain, and France. The link is 6500 km in length and carries 40,000 conversations per fiber, which is a significant improvement in the current copper cable, which carries only 200 conversations. The cable contains six single-mode fibers, two pairs operating continuously, while the third pair is for redundant usage. Repeaters are used every 51 km to regenerate the signal. The TAT-8 powers the repeaters with a copper central member that uses a 1.6 A current with 7500 V. This will not be the only oceanic link; soon all of the Pacific will be linked, from California to Hawaii to Japan, Australia, and Guam.

The automobile industry is always trying to reduce the cost, weight, and electromagnetic interference that occurs in cars. Fiber optic cables will be used to light display panels and indicate lamp failure. They will also be used to simplify wiring harnesses that transmit signals throughout the car. Plastic fiber will be used for this application.

The military uses the aspect of security to benefit from the fiber optic technology. Fiber optics has been used in avionics, submarines, ships, satellite earth stations, and transmission lines for tactical command post communications. The Navy's Airborne Light Optical Fiber Technology (ALOFT) connected 115 signals from the navigation and weapons system onto 13 fiber optic data channels. Before, 302 wires were used to multiplex all the signals. This was one of the earliest usages of fiber optics in the military. All 5000 communications and data-processing nodes of the intercontinental ballistic missiles will be connected together via fiber optics. Ships and submarines use fiber optics to reduce the shock, fire, and spark hazards that are always a concern aboard those vessels. Ground stations that link satellites use fiber optic cable for secure lines. The tactical command post likes the idea of fiber because of the weight reduction and easy deployment of the fiber optic cable.

The television industry uses fiber optics for short transmission links to span the gap from studio to transmitter, or live event to equipment van. The lighter weight of the fiber enables the camera operator to have a wider range of movement with the minicamera than with the conventional camera. The signals are modulated in analog form and transmitted baseband, requiring one channel, on a fiber optic cable. Six megahertz is sufficient for the bandwidth.

Surveillance and remote monitoring systems use fiber optics rather than copper because of the EMI and security advantages. The monitoring systems need only transmit a black-and-white picture, which means that each camera transmits only a single baseband analog modulated signal. The baseband signal modulates the source. A simplex or one-way transmission is sufficient for the monitoring stations for the railroad, parking lots, or military installations. If this type of transmission is not acceptable, a two-way or full-duplex system is needed and more complex equipment will have to be added.

The computer industry uses fiber optics to transmit digital information from one source (computers mainframe) to other buildings on the same campus. If the interconnections go beyond room to room and other buildings, intersite networks are required. Local area networks (LANs) are used for linking equipment in a building or distances of less than 1 km, metropolitan area networks (MANs) are used for linking services in a city, and wide area networks (WANs) are used for linking cities in the same area, such as Los Angeles to San Diego.

The end result of all of the applications will be a completely wired city. The fiber optic link will carry voice, video, and data which link computers, TV, cable television, shopping information, and local events and timetables to the home consumer. This will be built largely on a MAN, which will link all the above with factories and service organizations to provide the user with a broad range of services. A community in Japan is an example. The Higashi–Ikoma Optical Visual Information System, (HI-OVIS) an interactive system that a subscriber receives on multiple-channeled cable TV, was begun in 1976. The subscriber must have a TV set, keyboard, microphone, and printer. The system consists of 400 km of fiber optic cable serving 168 subscribers. HI-OVIS broadcasts on a subscriber network, providing such programs as shopping information, home study courses, local events, train timetables, and the local fire and police information.

One unique use of fiber optics is a fiber gyroscope, a long fiber coil with an optic signal traveling through it in both directions. The phase difference between the two signals is measured: Zero indicates no movement; having a phase difference means a rotation rate, which determines yaw, pitch, and roll.

The medical field has seen a surge in instrumentation that uses fiber optic cable. In 1987, F&P Fiber Optics AG of Switzerland introduced a microendoscope. The endoscope is used to inspect internal organs such as the lungs, colon, and intestines. The endoscope is 1.5 to 2.0 mm wide and 260 mm long, a lot smaller than earlier models. Not only does the small diameter help in reaching places in the human body, but there is less trauma to the system. Hospital stays for the patients are reduced, thus saving time and money.

There is some competition for fiber links, especially with transcontinental phone channels. The satellite cannot be beaten in connecting reliably to hard-to-reach places. Some remote places do not generate enough communications traffic to justify the cost of laying a cable. A satellite dish is a much cheaper way to carry the signal. When the TAT-8 cable was caught in a shark feeding frenzy, telephone conversations were automatically uplinked to a satellite, so service would be uninterrupted. Broadcasting one signal to many points is another advantage of using satellites. There has been talk in the fiber optic industry that instead of fiber to the home, cheap (less than $1000) satellite dishes will eventually win. Fiber optic cable can beat satellite transmission when there is heavy phone and data traffic. The fiber can hold more signals and has less noise. The satellite transmission takes a quarter of a second to make the round-trip 37,000-km geosynchronous orbit. This delay can cause an echo on the received end of the call. Satellite transmission can also be intercepted by spy satellites. Fiber does not have delay or security problems.

Whatever the future holds, there will be many new challenges to the fiber optic designer. Fiber optics will be a major player in the information revolution.

1.5 SUMMARY

Early optical communication systems sent light through air. Today, light is sent through glass smaller than a human hair. Not only is the small size attractive to various industries that use fiber optic cables, but the high bandwidth, no crosstalk, and security leave twisted pair and copper wire behind. The basic fiber system has an encoder for the signal and a light source or transmitter to convert the electrical to light. The optical fiber is the transmission medium that links to a light detector and amplifier.

Optical fibers are best suited for long-distance, high-bandwidth communications between two points. Applications of fiber optics can be found in many areas of our daily lives. The telephone industry has promoted and used fiber optics to link over 60% of global telecommunications. Television, the military, medical care, computers, and eventually our homes will reap the benefits that fiber optics has to give.

1.6 EQUATION SUMMARY

Total bandwidth for television broadcast:

$$BW_{video} = (analog\ bandwidth)(sampling\ rate)(number\ of\ encoded\ bits)$$

$$BW_{sound} = (sound\ bandwidth)(sampling\ rate)(number\ of\ encoded\ bits)$$

$$Total\ BW = BW_{video} + BW_{sound}$$

Number of calls per cable:

$$number\ of\ calls = (numbers\ of\ cables)(number\ of\ voice\ channels)$$

QUESTIONS

1. What is the structure of an optical cable?
2. What is the carrier for optical transmission if the carrier for electronic communications is the electromagnetic wave?
3. What are the most important reasons to use fiber optics?
4. What is the major reason that fiber optics is used underwater?
5. Can more than one signal be transmitted on a fiber optic cable?

PROBLEMS

1. The telephone transmission rate at the T4 level is 274.2 Mbps. Each phone message utilizes 64,000 bps. How many simultaneous messages could be sent along this system?

2. How many more calls (as a percentage) can the fiber optic cable carry than the twisted pair in Example 1-2?

3. If RG-19/U coaxial cable weighs 1110 kg/km and a fiber optic cable design weighs 6 kg/km, how much does 10 miles of each weigh?

4. If a commercial television broadcast were made up of a data rate of 96 Mbps and a sound track of 240 kbps, on what standard T-carrier line could this signal be transmitted?

5. Some fiber optic links operate at 20 gigabits per second (Gbps). If voice transmission is 64 kbps, how many voice channels can theoretically be carried on a single fiber?

2

The Principles of Light

CHAPTER OBJECTIVES

The student will be able to:

- Discuss the electromagnetic spectrum of which light is a part.
- Know the wavelengths that are important to fiber optic communications.
- Discuss and calculate the velocity of light in free space and in other materials.
- Know and use Snell's law.
- Describe and calculate the angles associated with total internal reflection (such as the critical angle and the acceptance angle).

2.1 INTRODUCTION

To understand fiber optic technology, the basics of light must be understood. This chapter will introduce these basics through the study of geometrical ray optics and Snell's law. Snell's law is important to the understanding of the operation of fiber optic cables.

2.2 ELECTROMAGNETIC SPECTRUM

Electromagnetic wave theory describes how light is propagated. Light is a type of electromagnetic wave and is classified by wavelength in the electro-

TABLE 2-1 Units of Measure

Frequency Designations		
10^{12} Hz	terahertz	THz
10^9 Hz	gigahertz	GHz
10^6 Hz	megahertz	MHz
10^3 Hz	kilohertz	kHz

Wavelength Designations		
10^{-6} m	micrometer	μm
10^{-9} m	nanometer	nm
10^{-15} m	femtometer	fm

magnetic spectrum. Table 2-1 shows the units of measure for frequencies and wavelengths.

The designations of the common frequencies in the electromagnetic spectrum are shown in Table 2-2. Light is often measured in terms of its wavelength, with units in nanometers (nm) or microns (μm). The preferred unit of measure is nanometers. Visible light has a narrow spectrum from 380 (violet) to 780 (red) nm. Optical communications use the wavelength range of the near-infrared, which is from 800 to 1600 nm. Most components for a fiber optic system operate more efficiently in these regions. The most useful of these wavelengths for optical waveguides are 850, 1300, and 1550 nm.

TABLE 2-2 Frequency Designations

Subsonic	0 to 10 Hz
Sound	15 to 2×10^4 Hz
Radio frequencies	10^3 to 10^9 Hz
AM, FM, shortwave, television	
Microwaves	10^9 to 3×10^{11} Hz
Infrared	3×10^{11} to 5×10^{14} Hz
Visible light	5×10^{14} to 7.7×10^{14} Hz
Red	384 to 482 THz
Orange	482 to 503 THz
Yellow	503 to 520 THz
Green	520 to 610 THz
Blue	610 to 659 THz
Violet	659 to 769 THz
Ultraviolet	7.7×10^{14} to 3×10^{17} Hz
X-ray	3×10^{17} to 5×10^{19} Hz
Gamma rays	10^{19} to 10^{21} Hz
Cosmic rays	10^{22} Hz

2.3 LIGHT AS A WAVE

Light has a constant speed of approximately 300,000,000 (3×10^8) meters per second (m/s), or 186,000 miles per second, in free space. Light is electromagnetic in nature, just like microwaves or other radio-frequency waves but at shorter wavelengths and higher frequencies. It is usually described as a wave, or more exactly a transverse wave, where the electric and magnetic field oscillate perpendicular to each other. The electric field effects the optical system more than does the magnetic field. The light beam is said to be *polarized* in the direction of oscillation of the electric field. If the magnetic or electric field follows a straight line, such a wave is said to be *linearly polarized*. If the field follows a circular or elliptical path, the wave is circularly or elliptically polarized. The polarization type is a consideration for designing lensed fiber optic system.

The wavelength is a sine wave moving in a z direction. A point on the wave moves a distance in time T. The velocity of light in free space is c (3×10^8 m/s). The distance traveled is the wavelength, defined as $T \times c$. Sine waves are expressed in frequency. Time is the inverse of frequency. The equations look like this:

$$\lambda = T \cdot c \qquad (2\text{-}1)$$

and

$$T = \frac{1}{f} \qquad (2\text{-}2)$$

Then

$$\lambda = \frac{c}{f} \qquad (2\text{-}3)$$

λ is the wavelength

c is the speed of light

f is frequency

Example 2-1

What is the wavelength in free space if the frequency is 0.45×10^{15} Hz?

<u>SOLUTION</u>

$$\lambda = \frac{c}{f}$$
$$= \frac{3 \times 10^8}{0.45 \times 10^{15}}$$
$$= 666.66 \times 10^{-9}$$
$$= 666.66 \text{ nm}$$

2.4 LIGHT AS A PARTICLE

Light is usually described as a wave, but as learned in physics, light has a dual nature. It can also be thought of as a stream of particles known as photons. The *photon* is a discrete packet or bundle of energy. The amount of energy the photon possesses is determined by the frequency of the wave: The higher the energy, the higher the frequency. This is an important idea, because it provides insight into how a *light emitting diode* (LED) works. When a LED emits light, electrons in the diode give up energy to produce photons, or light packets. Electrons in a LED can give up different amounts of energy. This results in different wavelengths, although the light is coming from the same LED. In contrast, when a laser emits light, all the electrons give up nearly the same amount of energy, and a single wavelength results. The energy of a single photon is given by the equation

$$E = h \cdot f = \frac{h \cdot c}{\lambda} \quad \text{joules} \qquad (2\text{-}4)$$

E is the energy of the photon, in joules (J)

h is Planck's constant 6.623×10^{-34}, in joule-seconds

f is the frequency, in hertz

c is the speed of light, in meters/second

Example 2-2

A LED source emits yellow light at a wavelength of 580 nm. What is the energy of one photon?

SOLUTION

$$E = \frac{h \cdot c}{\lambda} = \frac{(6.623 \times 10^{-34}) \cdot (3 \times 10^{8})}{580 \times 10^{-9}}$$
$$= 3.42 \times 10^{-19} \text{ J}$$

E is the energy of the photon, in joules

h is Planck's constant, in joule-seconds

c is the speed of light, in meters/second

λ is the wavelength, in meters

The energy of a photon is not as significant as the number of photons (N) emerging from a source. The value N shows how many photons could be emitted into the fiber end. The equation is

$$N = \frac{E \cdot \lambda}{h \cdot c}$$ (2-5)

N is the number of photons

E is the energy of the photon, in joules

λ is the wavelength, in meters

h is Planck's constant, in joule-seconds

c is the speed of light, in meters/second

Example 2-3

A LED source emits 250 watts (W) of yellow light at a wavelength of 580 nm. How many photons per second are emerging from the source?

SOLUTION

$$E = (250 \text{ watts}) \cdot (1 \text{ second})$$
$$= 250 \text{ watt-seconds} \quad \text{or} \quad 250 \text{ joules}$$
$$N = \frac{E \cdot \lambda}{h \cdot c} = \frac{(250) \cdot (580 \times 10^{-9})}{(6.623 \times 10^{-34}) \cdot (3 \times 10^{8})}$$
$$= 7.29 \times 10^{20} \text{ photons}$$

2.5 THE SPEED OF LIGHT

When given as 299,998 km/s, the speed of light is really the speed of light traveling through free space. However, when light enters a medium from free space, it changes speed and is bent or refracted, as seen in Figure 2-1. The factor by which the light changes speed is the *refractive index* or *index of refraction*. The index of refraction (n) of any optical medium is defined as the ratio between the speed of light in free space, c, and the speed of light in the medium, v. The refractive index is always greater than 1, varying with the color of light. The shorter the wavelength, the higher the refractive index.

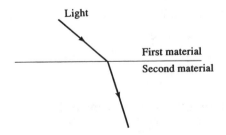

Light

First material

Second material

Figure 2-1 Light passage.

TABLE 2-3 Refractive Indices

Material	Refractive Index, n	Velocity of Light, v ($\times 10^6$ m/s)
Vacuum	1.000	300
Air	1.0003	299.9
Water	1.33	225.56
Glass	1.46–1.96	205.48–153
Diamond	2.42	123.97
Silicon	3.4	88.23
Gallium arsenide	3.6	83.33

Algebraically the equation is

$$n = \frac{c}{v} \tag{2-6}$$

Some common refractive indices are listed in Table 2-3.

2.6 INTERNAL AND EXTERNAL REFLECTIONS

The study of ray optics states that light rays show direction. If there are no obstacles and the light propagates in a homogeneous medium, the light rays will be straight lines. Rays follow the direction of the energy flow of the light beam. The speed of light changes as it enters a medium; therefore, it must be reflected or refracted in some way. The most practical case of refraction is that of a fisherman, looking at a fish in the pond from his boat. The fish is not exactly where the fisherman thinks it is, and the fisherman must compensate by putting the rod in at a different angle.

When light rays travel across two types of media, such as air and water, their direction of travel is altered, either by refraction or reflection. In the case of fiber optics, light is refracted from some light source into the cable end and then reflected down the cable. The cable is composed of more than one type of glass or plastic. There are three important cases that define the type of reflections that can be obtained when the light goes from one type of material to another.

There is a correlation between the refractive indices of two media and the path the light will follow. The law of reflection states that the plane of incidence contains the incident or incoming ray. The normal and reflected ray lies in the plane of incidence, and the reflected angle equals the incident angle. In Figure 2-2 the normal is an imaginary line perpendicular to the interface between the two media. The angle of incidence is the angle between the light in the first medium and the normal. The angle of refraction is the angle between the light in the second medium and the normal. Light can be reflected and refracted at the same time.

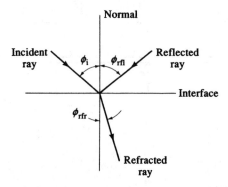

Figure 2-2 Refracted and reflected rays.

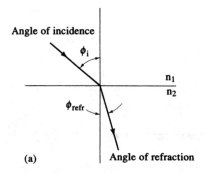

(a)

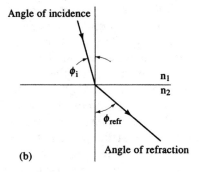

(b)

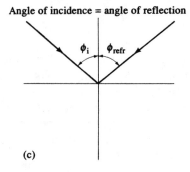

(c)

Figure 2-3 (a) Internal reflection, $n_1 < n_2$; (b) external reflection, $n_1 > n_2$; (c) total internal reflection, $n_1 \gg n_2$. (Courtesy of AMP, Inc.)

In Figure 2-3a, the light is passing from a medium of lower refractive index to a higher one, thus making the light bend toward the normal. This is called *external reflection*. In Figure 2-3b, the light is passing from a medium of higher refractive index to a lower-index medium. When the light is bent away from the normal, this is *internal reflection*. As the angle of incidence increases, the angle of refraction approaches 90°. When the angle of refraction is exactly 90°, the light does not enter the second medium but is reflected along the interface. The angle of incidence when this occurs is known as the *critical angle*. As the angle of incidence increases past the critical angle, the light is reflected at the interface and does not enter the second medium. The angle between the reflected light and the normal is the *angle of reflection*. Actually, the angle of reflection is equal to the angle of incidence as long as the angle of incidence is greater than the critical angle. This is *total internal reflection*. The light rays are reflected off the interface back into the denser medium. This is depicted in Figure 2-3c.

2.7 SNELL'S LAW

The law of refraction states that when the angle of incidence does not equal the angle of reflection, the light is bent as it enters the second medium. When light is refracted in the second medium, the angle of incidence is less than or equal to the critical angle. A relationship exists between the refractive indices of the two media, n_1 and n_2, and the angle of incidence and refraction, θ_1 and θ_2. This relationship, known as *Snell's law* or the *law of refraction*, is defined algebraically as

$$\frac{\sin \theta_1}{\sin \theta_2} = \frac{n_2}{n_1} \qquad (2\text{-}7)$$

Example 2-4

If a fisherman is looking at a fish at an angle of 30° in the water, find the angle at which the fish is located.

SOLUTION

$$n_1 = 1.0$$
$$n_2 = 1.333$$
$$\frac{\sin 30°}{\sin \theta_2} = \frac{1.3}{1.0}$$
$$\sin \theta_2 = \frac{\sin 30°}{1.3} = 0.3846$$
$$\theta_2 = \sin^{-1} 0.3846$$
$$\theta_2 = 22.62°$$

Snell's law is an important aspect of fiber optics. It is one of the theories behind the propagation of light along a fiber. To make the light travel down the fiber, the angle of incidence has to be greater than the critical angle. The critical angle occurs when the angle of incidence at which the transmitted ray is reflected along the surface of the interface is greater than the angle of the incoming ray. From Snell's law the following condition can be determined: $\theta_1 = \theta_c$ when θ_2 becomes 90° and $\sin \theta_2 = 1$:

$$\sin \theta_c = \frac{n_2}{n_1} \qquad (2\text{-}8a)$$

or

$$\theta_c = \sin^{-1} \frac{n_2}{n_1} \qquad (2\text{-}8b)$$

θ_c is the critical angle measured from the normal

n_2 is the refractive index of the cladding

n_1 is the refractive index of the core

For the critical case, $\theta = 90° - \theta_c$. For all angles less than the critical, total internal reflection takes place. Thus if the critical angle is known, the ratio of refractive indices is also known. This provides a value needed to decide what types of materials will become the core or the cladding.

2.8 OPTICAL FIBER

For simplicity, light can be treated as a ray. Think of discrete packets of light, propagating through the waveguide. Most of light rays making up these packets will be guided through the waveguide within the limits of its physical boundaries. The path the light will follow is a straight line, bouncing off the sides of the waveguide as a result of reflection. Some of the rays will get lost in the waveguide, but most will continue to reflect.

As shown in Figure 2-4, the fiber waveguide is constructed of two layers of glass or plastic. The inner layer, or the core, is where the light travels. The core material has a refractive index of n_1. The core is surrounded by an outer layer called the *cladding*. The cladding has a refractive index of n_2. These layers are protected by a jacket, or coating. The core has a higher refractive

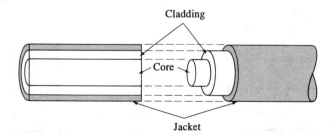

Cladding

Core

Jacket

Figure 2-4 Typical cable cross section.

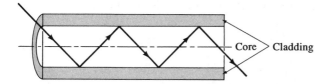

Core > Cladding

Figure 2-5 How light travels through a cable.

index than the cladding, which results in total internal reflection. This will occur only when $n_{\text{core}} > n_{\text{cladding}}$ (Figure 2-5). Light is injected or "launched" into the core at an incidence angle greater than the critical angle, striking the core–cladding interface. From Snell's law, when the angle of incidence is greater than critical angle, light is totally reflected. Since the angle of incidence is equal to the angle of reflection, the light will continue to travel down the fiber cable by total internal reflection. Any light striking the interface at less than the critical, that is, not within a region called the *acceptance cone*, will be absorbed or lost into the cladding (Figure 2-6). A large acceptance

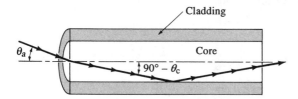

Figure 2-6 Acceptance cones.

cone allows the fiber to receive and propagate light from a LED, which has a larger field of light. The problem is that a lot of dispersion or pulse spreading is created. A narrow acceptance cone requires a narrower source of light, such as a laser. The cone of rays that can be accepted by the fiber is determined by the difference in the refractive indices. The difference, called the *fractional reflective index*, is given by the equation

$$\Delta = \frac{n_{\text{core}} - n_{\text{cladding}}}{n_{\text{core}}} \tag{2-9}$$

From the equation for the critical angle, all light rays that are not greater than $90 - \theta_c$ will be guided or transmitted in the core glass. For the light to be guided in the core, the light must be launched into the fiber from the outside. What would be the proper launch angle? The core of the fiber is very narrow, 50 μm or less. The launch angle between the light ray and the fiber axis can be derived by using the law of refraction:

$$\frac{\sin \theta_1}{\sin (90° - \theta_c)} = \frac{n_2}{n_1} \tag{2-10}$$

Since the refractive index of air is 1, n_1 is equal to 1 and the equation becomes

$$\sin \Theta = n_2 \sqrt{1 - \sin^2\theta_c}$$

With the equation above and the critical angle formula, the following will give us the acceptance angle Θ. The acceptance angle is the greatest possible angle that can be launched in to the core and still be guided in the optical waveguide:

$$\sin \Theta = \sqrt{n_1^2 - n_2^2} \qquad (2\text{-}11)$$

n_1 is the refractive index of the core

n_2 is the refractive index of the cladding

The acceptance angle is dependent only on the two refractive indices of the core and cladding. It is also known as the *acceptance cone half-angle*.

Example 2-5

A light ray enters from air to a fiber. The index of refraction of air is equal to 1. The fiber has an index of refraction of the core equal to 1.5. The index of refraction of the cladding is 1.48. Find the critical angle, the fractional reflective index, and the acceptance angle.

SOLUTION

Critical angle:

$$\sin \theta_c = \frac{n_2}{n_1} = \frac{1.48}{1.5}$$

$$\theta_c = \sin^{-1} \frac{1.48}{1.5}$$

$$= 80.63°$$

$$\theta = 90° - 80.63 = 9.37°$$

Total internal reflection will occur for all angles of less than 9.4°.

Fractional reflective index:

$$\Delta = \frac{n_{\text{core}} - n_{\text{cladding}}}{n_{\text{core}}} = \frac{1.5 - 1.48}{1.5}$$

$$= 0.013 \quad \text{or} \quad 1.3\% \text{ of the light}$$

Acceptance angle:

$$\sin \theta = \sqrt{(1.5)^2 - (1.48)^2}$$

$$= 0.244$$

$$\theta = 14.13°$$

2.9 NUMERICAL APERTURE

The sine of the acceptance angle is the *numerical aperture* (NA), which is the light-gathering ability of an optical fiber. The larger the NA, the greater the amount of light that can be accepted into the fiber, hence the greater the transmission distance that can be achieved. But if the NA is too great, the bandwidth of the system degrades. A warning about the NA of a fiber: The measured NA will be different from the material NA because a portion of the light is carried in the cladding.

$$NA = \sin \Theta \qquad (2\text{-}12)$$

Example 2-6

From Example 2-5, calculate the numerical aperture of the fiber.

SOLUTION

$$NA = \sin 14.13° = 0.244$$

This is a graded-index fiber, with an acceptable light-gathering ability.

The NA value is always less than 1. Typical values are 0.11 for a single-mode fiber, 0.21 for graded-index fibers, and 0.5 for plastic. The NA for a single-mode fiber is rarely specified. Light is not reflected or refracted in a single-mode fiber; therefore, there is no acceptance angle into the fiber.

2.10 FRESNEL REFLECTIONS

As light passes from one index of refraction to another, a small portion of the light is reflected back into the first material. The amounts of light reflected back are known as *Fresnel reflections*. Fresnel reflections at a boundary are defined by the equation

$$\rho = \left(\frac{n_2 - n_1}{n_2 + n_1}\right)^2 \qquad (2\text{-}13)$$

n_1 is equal to the index of refraction of the first material

n_2 is equal to the index of refraction of the second material

The light loss calculated in decibels (dB) is as follows:

$$dB = 10 \log_{10}(1 - \rho) \qquad (2\text{-}14)$$

Example 2-7

A light ray is entering the fiber cable from air, which has a refractive index of 1, to the core of the fiber, which has a refractive index of 1.5. What will be the Fresnel reflections, and how many decibels are lost?

SOLUTION

$$\rho = \left(\frac{1.5 - 1}{1.5 + 1}\right)^2 = 0.04$$

$$dB = 10 \log_{10}(1 - 0.04) = -0.177dB$$

This means that 4% of the light is reflected and 96% of the light is transmitted. Relatively little light is lost.

2.11 SUMMARY

The electromagnetic spectrum has designations for certain frequencies. Fiber optic communications use the wavelengths in the near-infrared, 800, 1300, and 1550 nm.

Light has a dual nature, either as a wave or as a particle known as a photon. As a wave, the wavelength of the light can be calculated. As a particle, the energy of the photon and the number of photons being emitted from a source can be found.

Light changes speed and is refracted as it enters different materials because of a physical property known as the refractive index. By using refraction, reflection, and Snell's law, the propagation of light can be described. This is called total internal reflection.

Light enters the fiber at or within twice the acceptance angle. The critical angle is the angle at which the light bounces off the core cladding interface. The numerical aperture (NA), the light-gathering ability of the fiber, has a range of 0.2 to 0.5. Fresnel reflections are a portion of light that is reflected back at the air–core interface.

2.12 EQUATION SUMMARY

Relation among wavelength, time, and speed of light:

$$\lambda = T \cdot c \tag{2-1}$$

Time–frequency relationship:

$$T = \frac{1}{f} \tag{2-2}$$

Relation among wavelength, frequency, and speed of light:

$$\lambda = \frac{c}{f} \tag{2-3}$$

Energy of a photon:

$$E = h \cdot f = \frac{h \cdot c}{\lambda} \quad \text{joules} \tag{2-4}$$

Number of photons emitted:

$$N = \frac{E \cdot \lambda}{h \cdot c} \tag{2-5}$$

Refractive index:

$$n = \frac{c}{v} \tag{2-6}$$

Snell's law:

$$\frac{\sin \theta_1}{\sin \theta_2} = \frac{n_2}{n_1} \tag{2-7}$$

Critical angle:

$$\sin \theta_c = \frac{n_2}{n_1} \tag{2-8a}$$

$$\theta_c = \sin^{-1} \frac{n_2}{n_1} \tag{2-8b}$$

Fractional reflective index:

$$\Delta = \frac{n_{\text{core}} - n_{\text{cladding}}}{n_{\text{core}}} \tag{2-9}$$

Law of refraction:

$$\frac{\sin \theta_1}{\sin (90° - \theta_c)} = \frac{n_2}{n_1} \tag{2-10}$$

Acceptance angle:

$$\sin \Theta = \sqrt{n_1^2 - n_2^2} \tag{2-11}$$

Numerical aperture:

$$\text{NA} = \sin \Theta \tag{2-12}$$

Fresnel reflections:

$$\rho = \left(\frac{n_2 - n_1}{n_2 + n_1} \right)^2 \tag{2-13}$$

Fresnel reflections:

$$dB = 10 \log_{10}(1 - \rho) \tag{2-14}$$

QUESTIONS

1. Explain what a photon is.
2. The wavelengths most commonly used for fiber optic transmissions are 800, 1300, and 1550 nm. To what part of the electromagnetic spectrum do they belong?
3. Is the velocity of light higher in water or in air?
4. Is there a difference between light traveling in free space (vacuum) and in air? Why?
5. What is the difference between refraction and reflection?
6. What is the numerical aperture?
7. How are the critical angle, acceptance angle, and Fresnel reflections related?

PROBLEMS

1. Calculate the wavelength in meters of the following.
   a. Far infrared signal, 1000 GHz
   b. Orange light
   c. Sound wave of 3 kHz
2. Find the velocity of light in the following materials.
   a. Diamond
   b. Fused silica, $n = 1.46$
   c. Glass, $n = 1.96$
   d. Sapphire, $n = 1.8$
3. A LED source emits 55 μW of red (780 nm) light. How many photons emerge from the source every 3 minutes?
4. A green LED (577 nm) has a speed of propagation in air of 3×10^8 m. What is its frequency?
5. What is the speed of light of a typical glass fiber optic cable ($n = 1.52$)?
6. Prove that the formula

$$\frac{\sin \theta_1}{\sin (90° - \theta_c)} = \frac{n_2}{n_1}$$

equals

$$\sin \Theta = n_2 \sqrt{1 - \sin 2\theta_c}$$

when $n_1 = 1$.

7. If a light ray travels through glass with an index of refraction of 1.43, by what percent has the speed of light been reduced?
8. What is the critical angle between glass ($n_1 = 1.52$) and air ($n_2 = 1.0$)?

9. What is the NA of fiber cable whose core has a refractive index of $n_1 = 1.45$ and whose cladding $n_2 = 1.43$?

10. What would be the Fresnel reflections of Problem 8? How many decibels are lost?

11. Light has entered two sheets of glass from air at an angle of 20°, as shown. What is the exiting angle from the right-hand plate?

air	glass	glass
$n = 1$	$n = 1.57$	$n = 1.63$

12. What is the transmitting frequency of a HeNe laser with a wavelength of 0.6328 μm, and how many photons per second are being emitted?

3

Optical Fiber and Its Properties

The student will be able to:

- Distinguish between the types of fiber and identify them by the core/cladding ratio.
- Describe and calculate the different types of dispersion associated with the fiber cable.
- Calculate bandwidth and bit rate.
- Define the different types of absorption.
- Know the difference between electrical and optical bandwidth.
- Know about the various bending losses.

3.1 INTRODUCTION

Choosing the correct fiber optic cable is one of the most important choices to make in designing any fiber optic system. This chapter focuses on the type of fiber optic cables and losses associated in the physical makeup of the fiber.

3.2 BASIC FIBER CONSTRUCTION

As shown in Chapter 2, the fiber is made up of two concentric layers called the core and the cladding. The *core* carries the light. The *cladding*, which has a different refractive index, provides the necessary total internal reflection phenomenon to occur. The index of refraction of the cladding is generally 1% lower than that of the core. The index of refraction ranges from 1.458 at a wavelength of 600 nm to 1.4469 at a wavelength of 1300 nm.

3.3 PROPAGATION OF THE LIGHT

Light propagation through a fiber depends on three major factors: size of the fiber, material or chemical composition of the fiber, and how the light is injected or launched into the fiber.

3.3.1 Size of the Fiber

The sizes of the fiber have been standardized nationally and internationally. Table 3-1 is a list of the diameters. Figure 3-1 compares the diameter size. The fiber is relatively small compared to the size of a human hair, which is about 100 μm. To designate the diameters in literature, it is expressed as 8/125. The first number is the core diameter and the second number is the cladding diameter.

3.3.2 Material Classification

Three material types are used in fiber optic cables; glass, plastic-clad silica, and plastic fiber. The most popular and most widely used is glass fiber. The glass is made from silicon dioxide or fused quartz. The indices of refraction for the core and cladding are varied by adding impurities. Germanium and phosphorus are added to increase the index of refraction. Boron and fluoride to decrease the index of refraction. These fibers are used for high information rates. The numerical aperture is low for these fibers, which can result in large coupling losses to the light source. Core diameters are 50, 100, and 200 μm. Overall fiber losses are a few dB/km.

Plastic-clad silica (PCS) fibers have a glass core and a plastic cladding. These fibers are good for shorter lengths (few hundred meters) and medium

TABLE 3-1 Core and Cladding Fiber Diameters

Core (μm)	Cladding (μm)
8	125
62.5	125
85	125
100	140

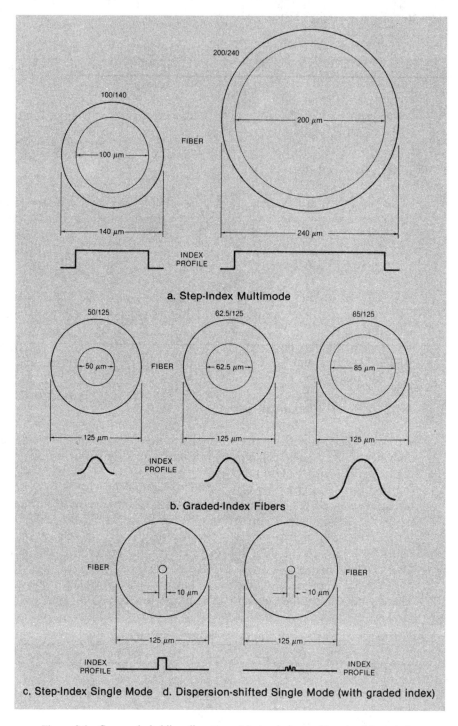

Figure 3-1 Core and cladding diameters: (a) step-index multimode; (b) graded-index fibers; (c) step-index single-mode; (d) dispersion-shifted single-mode (with graded index). (Courtesy of Sams: A Division of Macmillan Computer Publishing.)

TABLE 3-2 Fiber Types and NA Values

Construction	n_1	n_2	NA
Glass	1.48	1.46	0.24
PCS	1.46	1.40	0.41
Plastic	1.49	1.39	0.53

information rates. The numerical aperture is large but there are high losses. Core diameters are typically 200 μm. The fiber losses are 8 dB/km.

Plastic fibers have both a plastic core and cladding. Typical applications are very short runs and demonstration applications. There are three standard sizes for plastic fiber: 1000, 500, and 250 μm. The core diameters are 980, 490, and 240 μm, respectively. It is easy to see that plastic fiber has a large core and a small diameter. The fiber losses are several hundred dB/km.

Typical fiber types and numerical apertures are given in Table 3-2.

3.3.3 Launching the Light

The idea of total internal reflection was discussed in Chapter 2 using a ray model according to the laws of geometrical optics. In general, this model represents a good approximation to what is happening in the fiber, but not all fibers have the same core diameters. Core diameters may be as large as 1 mm or as small as 2 μm, which causes problems in the optical waveguide itself.

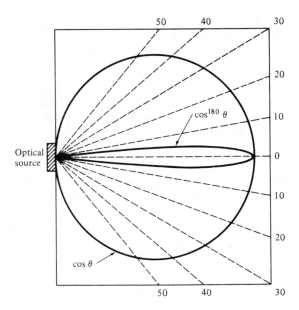

Figure 3-2 Diffused source.*

Given a more practical example of this concept, let's look at what happens when light is emitted from a small diffused source into the fiber cable.

Consider a small diffused source as depicted in Figure 3-2, known as a *Lambertian source*. The power that is radiated from this source on a sphere is known as the *far-field radiation intensity I* and is defined as follows:

$$I = I_0 \cos \phi \qquad \text{watts/steradian} \qquad (3\text{-}1)$$

I_0 is the maximum intensity

ϕ is the linear viewing angle

The total power Φ_0 emitted by the source is

$$\Phi_0 = \pi \cdot I_0 \qquad (3\text{-}2)$$

But the amount of power that can be collected by the fiber whose core diameter is greater than the source will be

$$\Phi = \Phi_0 \cdot (NA)^2 \qquad (3\text{-}3)$$

The fraction of optical power from a diffused source that a fiber will propagate is

$$\frac{\Phi}{\Phi_0} = (NA)^2 \approx 2 \cdot n \cdot \Delta n \qquad (3\text{-}4)$$

NA is the numerical aperture

n is the refractive index of the core

Δn is the difference between the core and cladding refractive index

To collect as much light as possible, it is necessary to make the core refractive index and the difference in the refractive index of the core and the cladding as large as possible. This is not a practical situation, because it would mean that to collect all of the light, a fiber would use glass of a high refractive index with no cladding. The reason we need a cladding is attributable to a phenomenon called an *evanescent wave*. It is an electromagnetic disturbance that does penetrate the reflecting surface. The evanescent wave usually does not propagate in a medium of lower refractive index, but decays exponentially as it moves away from the boundary surface. If the wave encounters any variation or nonuniformity at the boundary, the wave will propagate. Therefore, in an unclad fiber, the evanescent waves would become significant, because of the surface conditions or nonuniformity at the boundary of the fiber. The outside of the core cannot be controlled as well in the fiber-making process. It is worthy to note that the power that was launched into the fiber will be coupled out of the fiber and high attenuation will result.

Example 3-1

A step-index fiber is being used for a transmission system. The core has a refractive index of 1.5 and the cladding has a refractive index of 1.48.

Find the fraction of optical power from a diffused source that the fiber will propagate.

<u>SOLUTION</u>

$$NA = \sqrt{n_1^2 - n_2^2}$$
$$= \sqrt{(1.5)^2 - (1.48)^2}$$
$$= 0.244$$

The acceptance angle is $\sin^{-1}$ of 0.244, which is equal to 14.12°. Therefore, the fraction of optical power from a diffused source that the fiber will propagate is the square of the numerical aperture: $(NA)^2 = 0.06$, or 6% of the light. This may seem small, but there is no evanescent wave problem.

3.4 MODES AND THE FIBER

Fiber optic cable can also be called an *optical waveguide*. The energy propagation representing a light signal can be analyzed by using Maxwell's electromagnetic field equations, with applicable boundary conditions for cylindrical coordinates for a waveguide. These equations are analyzed by taking into account material constants, refractive indices, and the cylindrical boundaries of the core–cladding interfaces. It is found that the propagating energy has distinct sets of solutions known as *Bessel functions* or *modes*. The solutions to Maxwell's equation is in the form of a harmonic in space and time of the sine and cosine functions. The light signal is propagated down the waveguide in these different modes, which results in different propagation characteristics, such as velocity and wavelength. This causes some distortion of the light signal. When the light signal is received at the end of the waveguide, there will be degradation from the original wave shape.

An important quantity in describing how many modes a fiber can propagate is the *V number*. Although also called the *characteristic waveguide parameter, mode volume number*, or *normalized wave number*, the most common nomenclature is "*V* number." A dimensionless quantity, its equation is

$$V = 2\pi \cdot \frac{a}{\lambda} \cdot NA \qquad (3\text{-}5)$$

a is the core radius, in meters

λ is the wavelength, in meters

NA is the numerical aperture

If the *V* number of the fiber becomes smaller than 2.405, only a single-mode can propagate in the core. The number of modes in a step-index profile is approximately

$$N = \frac{V^2}{2} \tag{3-6}$$

The number of modes for a graded-index fiber is

$$N = \frac{V^2}{4} \tag{3-7}$$

Example 3-2

A graded-index fiber has a core diameter of 50 μm and a numerical aperture of 0.22 at a wavelength of 0.850 μm. What are the V number and the number of modes guided in the core?

SOLUTION

$$V = 2\pi \cdot \frac{50/2}{0.850} \cdot 0.22 = 40.66$$

$$N = \frac{(40.66)^2}{4} = 413.31$$

This type of fiber is called *multimode graded-index fiber*, for obvious reasons.

3.5 REFRACTIVE INDEX PROFILE

The propagation of the modes in a fiber depends on the shape of the refractive index profile. The refractive index profile is a function of the refractive index n as a function of the radius r. It is used to describe the change in the refractive index from center of the fiber's core out to the cladding.

There are four main types of refractive index profiles: graded, step, triangular, and parabolic. Only in the step profile is the refractive index constant in the core glass. In all others, the refractive index of the core glass rises gradually from n_2, the refractive index of the cladding, to n_1, the refractive index of the core at the axis. The triangular and, in particular, the parabolic profiles are called graded-index profiles. Figure 3-3 shows the most popular refractive index profiles.

3.6 TYPES OF FIBER

There are three basic types of fiber: multimode step index, graded-index multimode, and single-mode. All the fibers have advantages, costs, and applications associated with them. Table 3-3 summarizes these.

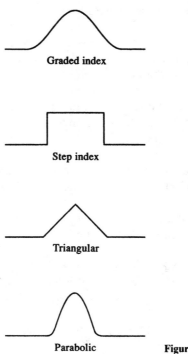

Graded index

Step index

Triangular

Parabolic

Figure 3-3 Refractive index profiles.

TABLE 3-3 Fiber Types

	SINGLE-MODE FIBER	GRADED-INDEX MULTIMODE FIBER	STEP-INDEX MULTIMODE FIBER
Cladding / Core / Protective Plastic Coating			
SOURCE	REQUIRES LASER	LASER or LED	LASER or LED
BANDWIDTH	VERY VERY LARGE > 3 GHz-km	VERY LARGE 200 MHz to 3 GHz-km	LARGE < 200 MHz-km
SPLICING	VERY DIFFICULT DUE TO SMALL CORE	DIFFICULT BUT DOABLE	DIFFICULT BUT DOABLE
EXAMPLE OF APPLICATION	SUBMARINE CABLE SYSTEM	TELEPHONE TRUNK BETWEEN CENTRAL OFFICES	DATA LINKS
COST	LESS EXPENSIVE	MOST EXPENSIVE	LEAST EXPENSIVE

3.6.1 Multimode Step-Index Fibers

The simplest type of fiber is called the step-index fiber. The multimode step-index fiber has a glass core 50 to 200 μm in diameter surrounded by a glass cladding. Large-core diameters result in more modes of light than can be accomplished with small-core diameters. This type of fiber has been mentioned before, with the refractive index of the cladding slightly lower than that of the core. The common term for this fiber is *multimode fiber with a step-index profile*. The ray path and the profile of the refractive index is shown in Figure 3-4.

As depicted in the figure, there are two rays that can travel along the core. One is called the *axial ray*, which travels along the axis. The other is called the *marginal* or *meridional ray*, which travels along a path near the critical angle. The meridional ray will travel farther than the axial ray, and the additional distance traveled is defined by

$$\Delta Z = \frac{Z \cdot (n_{core} - n_{cladding})}{n_{cladding}} \quad \text{meters} \tag{3-8}$$

The additional time it takes for the meridional ray to travel is

$$\Delta t = \frac{Z \cdot (n_{core} - n_{cladding}) \cdot n_{core}}{c} \quad \text{seconds} \tag{3-9}$$

Example 3-3

A step-index fiber has a refractive index of the core of 1.5 and a refractive index of the cladding of 1.48. For a 25 km link, what is the additional distance traveled for the meridional ray, and the additional time it will take?

SOLUTION

$$\Delta Z = \frac{(25 \times 10^3)(1.5 - 1.48)}{1.48}$$
$$= 337.84 \text{ m}$$
$$\Delta t = \frac{(25 \times 10^3)(1.5 - 1.48)(1.5)}{3 \times 10^8}$$
$$= 2.5 \ \mu s$$

This shows there will be a delay of 2.5 μs from the axial ray, thus causing modal dispersion.

This time delay, known as *modal dispersion*, causes distortion in the pulse that is being sent. It causes a short light pulse to broaden, thus reducing the transmission speed, or bit rate, and the transmission bandwidth. The

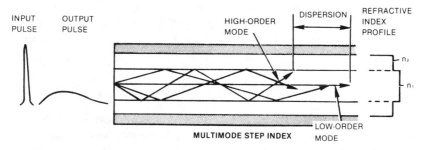

Figure 3-4 Step-index fiber and refractive index profile. (Courtesy of AMP, Inc.)

modes that the axial ray carries and the modes that the marginal ray carries interact with each other, exchanging energy along the way, causing mode mixing.

Modal dispersion is typically 15 to 30 nanoseconds per kilometer (ns/km). If the distance is doubled, the dispersion time will double. Although this seems insignificant at short distances, fiber optic systems can transmit data over much longer distances. Dispersion could limit the entire system's bandwidth. The dispersion can also be expressed in frequency, such as 100 MHz-km. This number indicates that the highest operating bandwidth is 100 MHz for a 1 km fiber before dispersion will be a limiting problem in the system.

Since there is a lot of dispersion associated with the multimode step-index fibers, it is the least efficient of the three types of fibers. However, the fiber is the least inexpensive, is easy to terminate, lends itself to the addition of end connectors, and has a large numerical aperture through which light can enter the fiber. These fibers are used for short runs, less than a kilometer, where the required signal bandwidths are smaller. Using a single-mode or graded-index mode fiber will lessen the pulse broadening.

3.6.2 Single-Mode Step-Index Fiber

Modal dispersion can be lessened by reducing the core's diameter until the fiber will transmit only one mode. The single-mode fiber or monomode fiber is based on this principle, having a core diameter of only 2 to 8 μm. Figure 3-5 shows the path of light and the refractive index profile. The numerical aperture, and therefore the acceptance angle, is small for these fibers, which makes launching the light more difficult. Only the fundamental mode can be

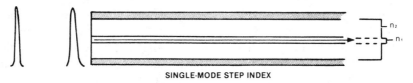

Figure 3-5 Single-mode fiber and refractive index profile. (Courtesy of AMP, Inc.)

used to transmit the energy in the fiber. These fibers are the most efficient but are difficult to work with because of their small core diameters, especially when it comes to splicing or terminating the fiber. Single-mode fiber is used for very high speed, large-bandwidth, long-distance applications.

3.6.3 Multimode Graded-Index Fiber

The multimode graded-index fiber is a compromise of the two fiber types described previously. The boundary between the core and cladding is not as sharply defined as with step-index fiber. The refractive index of the core glass diminishes parabolically to the interface between the core and the cladding. Figure 3-6 shows the ray paths and the refractive index profile.

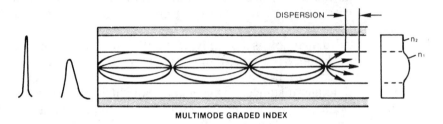

Figure 3-6 Multimode graded index and refractive index profile. (Courtesy of AMP, Inc.)

The light rays travel skew or helical patterns in a multimode graded-index fiber. Because of the parabolic refractive index profile, the rays are refracted continuously and change their propagation direction. The rays traveling on the fiber axis transverse a shorter path than the ones near the core–cladding interface. The difference in refractive index nullifies the time-delay problem that was encountered with the step-index fiber. The dispersion is reduced down to roughly 2 ns/km. Multimode graded-index fibers are approximately 125 μm in size. They are easily terminated but cost more than step-index fiber.

3.7 DISPERSION

The technical definition of *dispersion* is pulse broadening or spreading of the light as it travels down the optical fiber. This, in turn, affects the fiber bandwidth. Dispersion is divided into two principal categories: modal or multimode and chromatic dispersion. Chromatic dispersion is subdivided into material and waveguide dispersion. All the dispersion measurements can be characterized in the time domain (ns/km) or the frequency domain (MHz-km). The units of measure most often used are picoseconds per nanometer-kilometer (ps/nm-km). Although this seems rather lengthy, the units have separate meanings. The time measurement in picoseconds describes the in-

                                             Optical Fiber and Its Properties   *Chap. 3*

crease in the pulse width. Nanometers are the unit of measure for the pulse width of a typical light source, and kilometers represents the length of the cable.

3.7.1 Modal Dispersion

Modal dispersion is found in multimode fibers. The axial and meridional rays travel at different speeds, although their phase velocity is the same. The meridional ray transverses a longer path than the axial, causing a time delay. Therefore, a pulse transmitted into the fiber will propagate over several paths and be received at the end at slightly different times.

The axial ray has the shortest mode path to travel. The path is calculated by the formula

$$z_t(\text{min}) = \frac{z}{\sin \phi(\text{max})} = \frac{z}{\sin 90°} = z \quad \text{meters} \quad (3\text{-}10)$$

z is the total fiber length, in meters

ϕ is the angle of incidence of the mode and the fiber tip

$\phi(\text{max})$ is 90° for the lowest-order mode

This means that the minimum path length is the length of the fiber. The meridional ray has the longest path. Its path length is calculated by

$$z_t(\text{max}) = \frac{z}{\sin \phi(\text{min})} = \frac{z}{\sin \phi_c} = z \cdot \frac{n_{\text{core}}}{n_{\text{cladding}}} \quad \text{meters} \quad (3\text{-}11)$$

z is the total fiber length

ϕ is the angle of incidence of the mode and the fiber tip

$$\phi(\text{min}) = \sin^{-1} \left(\frac{n_{\text{cladding}}}{n_{\text{core}}} \right) \approx \frac{n_{\text{cladding}}}{n_{\text{core}}}$$

The total path length difference due to this modal dispersion is

$$\Delta_z = z_t(\text{max}) - z_t(\text{min}) = z \cdot \left(\frac{n_{\text{core}}}{n_{\text{cladding}}} - 1 \right) \quad \text{meters} \quad (3\text{-}12)$$

Often, a substitution is made for $n_{\text{core}}/n_{\text{cladding}}$ in the form

$$\Delta = \frac{n_{\text{core}} - n_{\text{cladding}}}{n_{\text{core}}} \quad (3\text{-}13)$$

The difference in path length then becomes

$$\Delta_z = z \cdot \frac{\Delta}{1 - \Delta} \quad \text{meters} \quad (3\text{-}14)$$

The total time delay due to modal dispersion is defined by dividing the difference in path length by the phase velocity. The phase velocity, v_p, is the speed of light divided by the refractive index of the core. The equation is

$$\Delta_{t\text{ modal}} = \frac{\Delta_z}{V_p} = \frac{n_{\text{core}} \cdot z}{c} \cdot \frac{\Delta}{1 - \Delta} \quad \text{ns} \qquad (3\text{-}15)$$

Example 3-4

For a step-index multimode fiber, determine the dispersion per kilometer of length and the total dispersion for a 20 km length. The index of refraction for the core is $n_1 = 1.48$, and the cladding has an index of 1.46.

SOLUTION

$$\Delta = \frac{1.48 - 1.46}{1.48} = 0.0135$$

$$\Delta t = \frac{1.48 \cdot 1000 \cdot 0.0135}{(3 \times 10^8) \cdot (1 - 0.0135)} = 6.75 \times 10^{-8}$$

$$= 67.5 \text{ ns}$$

For a 20 km length the total dispersion would be

$$\tau = (67.5 \text{ ns}) \cdot (20 \text{ km})$$
$$= 1.35 \text{ ms}$$

Multimode graded-index fibers can be made to have a lower modal dispersion than that for a multimode step-index fiber. The graded-index fibers have a parabolic refractive index profile. The light travels a zigzag path from a higher to a lower refractive index and back again. This causes the group velocity to rise and then go lower. The result is an averaging of the group velocity, thus lowering the total dispersion. The minimum value for a graded-index fiber is

$$\Delta_{t\text{ modal}} = \frac{n_{\text{core}} \cdot z \cdot \Delta^2}{8 \cdot c} \quad \text{seconds}$$

Practical limits for total modal dispersion are about 1 ns/km.

3.7.2 Material Dispersion

Material dispersion is found in single-mode fibers and multimode fibers. However, in multimode fibers, modal dispersion greatly exceeds the magnitude of the material dispersion. When a pulse of light is transmitted, it is centered about certain frequency, but there are components of this light that have several different frequencies. The shorter wavelengths will be delayed more than the longer wavelengths. This will cause dispersion at the receiver. Material dispersion is dependent on the dopants of the core glass.

These dopants can be varied to influence the "zero" point. The curve of material dispersion is a Gaussian curve that crosses zero at 1.3 μm. If the light source launched into the single-mode fiber is 1.3 μm, the dispersion can be canceled out.

Material dispersion can be calculated by taking the second derivative of the index of refraction equation with respect to wavelength. The formula is

$$\Delta_{t \text{ material}} = -\frac{z}{c} \cdot \lambda_0 \cdot \frac{d^2n}{d\lambda^2} \cdot \lambda_{3dB} \qquad \text{ps/nm-km}$$

z is the total length of the fiber

λ_0 is the center wavelength

λ_{3dB} is the spectral width of the light source, in nanometers

The total material dispersion is

$$\Delta_{t \text{ material}} = D_m \cdot z \cdot \lambda_{3dB} \qquad \text{seconds} \tag{3-16}$$

D_m is the dispersive coefficient based on the material, in ps/ns-km

z is the total length of the fiber, in kilometers

λ_{3dB} is the spectral width of the light source, in nanometers

Example 3-5

A single-mode doped silica fiber has a dispersive coefficient of 40 ps/nm-km at a center wavelength of 1.3 μm. The light source has a spectral width of 1.5 nm. Find the material dispersion if the length of the fiber is 20 km.

SOLUTION

$$\Delta_{t \text{ material}} = (40 \text{ ps/nm-km}) \cdot (1.5 \text{ nm}) \cdot (20 \text{ km}) = 1.2 \text{ ns}$$

Material dispersion limits the bandwidth for laser-based multimode fibers at 850 nm. At 1300 nm, modal dispersion increases for both LED and laser systems.

3.7.3 Waveguide Dispersion

Waveguide dispersion is important in single-mode fibers. The distribution of light from the fundamental mode in the core and cladding is wavelength dependent. The higher the wavelength, the more the fundamental mode will spread from the core into the cladding. The increase in light in the cladding will cause the fundamental mode to propagate faster. This causes a time delay to develop, and only an average for the velocity of propagation is

---

significant. Waveguide dispersion can be greatly influenced by changing the profile structure of the refractive index. The formula for calculating the waveguide dispersion is

$$\Delta_{t \text{ waveguide}} = \Delta_w \cdot \lambda_{3dB} \cdot z \qquad \text{seconds} \qquad (3\text{-}17)$$

Δ_w is the dispersive coefficient based on the material's theoretical limit; the value is 6.6 ps/nm-km

λ_{3dB} is the spectral width of the light source

z is the total length of the fiber

Example 3-6

A single-mode doped silica fiber has a dispersive coefficient of 6.6 ps/nm-km at a center wavelength of 1.3 μm. The light source has a spectral width of 1.5 nm. Find the waveguide dispersion if the length of the fiber is 20 km.

SOLUTION

$$\Delta_t = (6.6 \text{ ps}/\mu\text{m}) \cdot (1.5 \text{ nm}) \cdot (20 \text{ km}) = 198 \text{ ps}$$

The sum of material and waveguide dispersion is called *chromatic dispersion*.

3.7.4 Total Dispersion and Bit Rate

The net effect of all the dispersions will cause pulse broadening of the light. Total dispersion is the root mean square of all the dispersions:

$$\Delta_{t(\text{total})} = \sqrt{\Delta_{\text{modal}}^2 + \Delta_{\text{material}}^2 + \Delta_{\text{waveguide}}^2} \qquad \text{seconds} \qquad (3\text{-}18)$$

The pulse width of the received light is the sum of the transmitted pulse width and the total dispersion:

$$t_r = t_w + \Delta_{t(\text{total})} \qquad \text{seconds} \qquad (3\text{-}19)$$

The theoretical maximum bit rate for a digital receiver will be

$$B = \frac{1}{t_r} = \frac{1}{t_w + \Delta_{t(\text{total})}} \qquad (3\text{-}20)$$

Equation (3-20) is for an ideal situation; therefore, a safety factor will be built into the next equation. The bit rate depends on the transmission rate. This equation can be used only for a return-to-zero bit pattern (1101 . . .):

$$B = \frac{0.441}{t_r} = \frac{0.441}{t_w + \Delta_{t(\text{total})}} \qquad (3\text{-}21)$$

Note that these equations are good for a Lambertian source at the full width half maximum spectral width. Spectral width values for a laser diode are 3 to 5 nm. Light emitting diodes have spectral widths from 40 to 70 nm at 850 nm wavelength, and 120 to 150 nm at 1300 nm wavelength.

Example 3-7

The fiber in the previous examples will be used to determine the received pulse width and the maximum bit rate. A laser diode with a spectral width of 3.5 nm will be used.

SOLUTION

Total dispersion:
$$\Delta_{t(total)} = \sqrt{0.0 + (1.20 \times 10^{-9})^2 + (198 \times 10^{-12})^2}$$
$$= 1.216 \text{ ns}$$

Received pulse width:
$$t_r = t_2 + \Delta_{t(total)}$$
$$= 3.5 + 1.216 \text{ (ns)}$$
$$= 4.716 \text{ ns}$$

The bit rate for a digital receiver will be
$$B = \frac{0.441}{4.716 \text{ ns}} = 93.5 \text{ Mbps}$$

3.8 DATA RATE AND BANDWIDTH

The *data rate* is the rate at which information is transferred in bits per second. The *baud rate* is the number of signal transitions in a second. The baud rate is related to the data rate or the bit rate at which the system can transmit. In binary systems, the baud rate and the bit rate are interchangeable.

The *bandwidth* of a fiber is the range of frequencies that can be transmitted with minimal distortion. The maximum bit rate is approximately equal to the bandwidth. Bandwidth is used as a measure for the dispersion properties of an optical fiber. Because of the dispersion, the pulse that is transmitted is spread or broadened in time as it propagates down the fiber. The fiber in effect acts like a low-pass filter, letting low frequencies pass and attenuating the rest.

Single-mode fibers are specified by dispersion or bandwidth in picoseconds per nanometer-kilometer (ps/nm-km). The formula for finding the bandwidth is only an approximation for an analysis using transfer functions;

therefore, in a simplified form the equation becomes

$$BW = \frac{0.187}{\Delta_{t(total)} \cdot t_w \cdot z}$$

$\Delta_{t(total)}$ is the total dispersion at the operating wavelength

t_w is the spectral width (rms) of the source, in nanometers

z is the length of the fiber, in kilometers

Example 3-8

The dispersion coefficient of a single-mode fiber is 3.5 ps/nm-kilometer. The spectral width of the source is 1.8 nm and the length of the fiber is 50 km. Find the bandwidth.

SOLUTION

$$BW = \frac{0.187}{(3.5 \text{ ps/nm-km}) \cdot (1.8 \text{ nm}) \cdot (50 \text{ km})}$$
$$= 593.65 \text{ MHz}$$

The bandwidth–length product is used as a specification for multimode fiber in megahertz-kilometers. If the bandwidth–length product is 600 MHz-km, a signal of 600 MHz rms can be sent 1 km without dispersion affecting it.

Example 3-9

Find the bandwidth–length product if the bandwidth of the multimode fiber is 60 MHz and the length is 50 km.

SOLUTION

$$BWLP = (60 \times 10^6) \cdot (50 \times 10^3) = 3 \times 10^{12} = 3000 \text{ GHz-km}$$

There is a distinction between electrical and optical bandwidths. The *electrical bandwidth* is defined as the frequency at which the ratio between I_{out}/I_{in} drops to 0.707. The *optical bandwidth* is the frequency at which the power$_{out}$/power$_{in}$ drops to $\frac{1}{2}$. The optical bandwidth is larger than the electrical bandwidth, but the parameters represent two ways to describe the same system:

$$BW_{electrical} = 0.707 BW_{optic} \qquad (3-22)$$

This means that if the optical bandwidth of the fiber is 0.23 GHz-km, the electrical bandwidth is 0.16 GHz-km.

3.9 ATTENUATION

Attenuation gives the loss of light power along an optical fiber. Materials have to be chosen so that there is little loss in the frequency range of 0.5 and 1.6 μm. Figure 3-7 shows the total attenuation curve for a germanium oxide–doped silica fiber. Attenuation is classified into two principal groups: scattering and absorption.

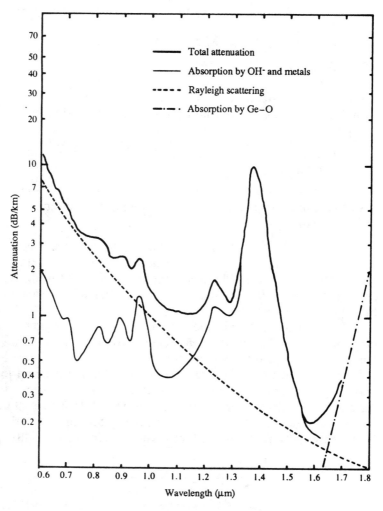

Figure 3-7 Attenuation for a germanium oxide–doped silica fiber. (Courtesy of Les Editions Le Griffon D'Argile.)

3.9.1 Scattering

Scattering is due to imperfections in the glass fiber as it is heated in the forming process. The dopants and microscopic variations become fixed in the glass, causing mirror-like reflections in the fiber as shown in Figure 3-8. This is known as *Rayleigh scattering* and is proportional to $1/\lambda^4$. This loss can be minimized by careful cooling of the fiber as it is drawn from the melt.

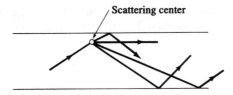

Scattering center

Figure 3-8   Scattering.

3.9.2 Absorption

Absorption is basically a material property, due primarily to electronic and atomic resonances in the glass structure. Absorption is divided into three groups:

1. *Ultraviolet absorption* has to do with the electronic structures of the crystal atoms.
2. *Infrared absorption* results from the lattice vibrations of the atoms.
3. *Ion resonance* is related to the amount of water content in the air as the fiber is forming.

Ultraviolet absorption is caused by the valence electrons in the pure silica fiber material to be ionized. When this happens, energy is lost, which in turn contributes to the light loss in the fiber. Values for ultraviolet absorption are approximately 0.1 dB/km. Doping the fiber with germanium oxide seems to reduce this parameter.

Infrared absorption is due to photons being absorbed by atoms within the glass material and converting that energy to atomical vibrations. The typical values for this are 0.5 dB/km.

Ion absorption or OH$^-$ absorption can be a significant loss in the fiber-making process because minute quantities of water molecules get trapped in the glass. The absorption peaks occur at 0.95, 1.25, and 1.39 μm. For the ion absorption peaks not to be spread out, the water content of the fiber has to be kept below 0.01 parts per million. The amount of humidity when the fiber

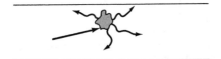

Figure 3-9   Absorption.

is forming must be closely regulated. Figure 3-9 gives the total absorption losses.

3.10 LEAKY MODES

As discussed earlier, light energy can be made up of meridional rays (the rays of propagation) and skew rays (the helically coiled light rays). If the skew rays' angle of incidence is greater than the critical angle, the axial and radial modes will propagate in the fiber. The radial component mode leaves the core, entering the cladding, causing the light power to be reduced. A technique known as *mode stripping* is used to reduce these leaky modes. A thin coating of low-purity glass that has a higher index of refraction is applied to the outside of the cladding. This causes only a small fraction of the light to stay in the cladding and not reenter the core to be transmitted. Mode coupling losses occur when the propagated mode and the leaky mode unite or couple due to small bubbles in the glass. Obviously, this must be avoided in the manufacturing process of the fiber.

3.11 BENDING LOSSES

Bending losses associated with the fiber are classified as microbending or macrobending, shown in Figure 3-10. These losses occur more frequently in the making and/or installation of the fiber. *Microbending* is caused by small

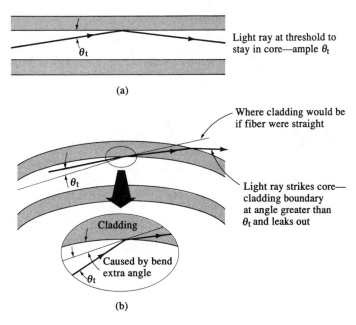

Figure 3-10 Bending losses: (a) straight fiber; (b) bent fiber. (Courtesy of Sams: A Division of Macmillan Computer Publishing.)

cracks in the glass. In the process of making the fiber, the core and cladding are joined and cooled together to form the total fiber. There are different thermal cooling rates for the core and the cladding, causing contractions or cracks in the glass. Mode coupling can occur at these cracks.

Macrobending occurs when the cracks run the length of the fiber rather than the width. During installation, corners and sags between poles put a lot of strain on the fiber. The tension variation on the fiber exceeds the allowable limit specified by the manufacturer; therefore, cracks appear in those areas, causing modes not to propagate.

3.12 CUTOFF WAVELENGTH

As seen earlier in this chapter, one mode or several modes may propagate in the fiber. Cutoff wavelength is used to describe the lowest operational wavelength in which only the fundamental mode can be guided. The fiber is considered multimode at wavelengths lower than the cutoff or λ_c and single-mode at higher wavelengths. The equation to calculate the cutoff frequency is

$$\lambda_c = \pi \cdot \frac{2 \cdot a}{V_c} \cdot \text{NA} \tag{3-23}$$

a is the core radius

V_c is the normalized frequency cutoff, in the range $1.6 < V < 3.5$

NA is the numerical aperture

The normalized frequency cutoff value for single-mode fiber is 2.405. This number is derived mathematically from a model of the waveguide (Maxwell's equations) and is a solution of the Bessel function. For graded-index fiber, the normalized frequency cutoff is 3.4.

Example 3-10

A single-mode fiber has a core diameter of 9 μm and a NA of 0.11. What is the cutoff wavelength?

SOLUTION

$$\lambda_c = \frac{(\pi) \cdot (9 \ \mu\text{m}) \cdot (0.11)}{2.405}$$
$$= 1293 \text{ nm}$$

Since the cutoff wavelength is 1293 nm, this means that a source transmitting at a wavelength below 1293 nm will cause the single-mode fiber to look like a multimode fiber.

3.13 MODE FIELD DIAMETER

The numerical aperture is used to describe the light-gathering ability of the fiber in just the core only. The *mode field diameter* is used to describe the light entering and propagating in the core and in the cladding. It is also known as the *spot size*. Figure 3-11 depicts the difference in size from the core diameter and the mode field diameter. This parameter is used to measure the spot size of light propagating down a single-mode fiber. For Gaussian power distributions, the spot size is the $1/e^2$ value of the peak intensity of the light. The mode field diameter, $2w_0$, has been defined to quantify the size of the fundamental mode or the radial field diameter. The mode field radius is where the fundamental mode decreases to 37%, or $1/e$ of its maximum optical fiber axis. This field radius is dependent on the wavelength, increasing as the wavelength increases. For a single-mode fiber the mode field radius can be approximated by the equation

$$w_0 = \frac{2.6}{V_c} \cdot a \qquad (3\text{-}24)$$

V_c is the normalized frequency

a is the core radius

Example 3-11

For Example 3-10, find the mode field radius.

SOLUTION

$$w_0 = \frac{(2.6) \cdot (4.5 \times 10^{-6})}{2.405} = 4.865 \times 10^{-6} \text{ m}$$

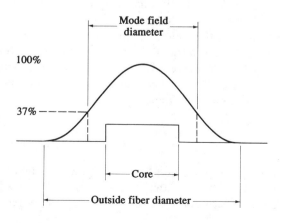

Figure 3-11 Mode field diameter.

The wavelengths that are important to the fiber are between 1150 and 1875 nm.

3.14 OTHER FIBER TYPES

In the past few years, new advances in fiber technology have made fibers that can be used for various applications. Plastic fiber and dispersion-shifted fiber have benefited different industries.

3.14.1 Plastic Fiber

Plastic fibers are multimode step-index fibers manufactured of polymethyl methacrylate (PMMA). The core of the fiber still has a refractive index higher than the cladding. The cladding is made of a fluorine-containing polymer. The size of the fiber can range from 1000 to 250 μm. The core size ranges from 980 to 230 μm.

Plastic fiber is very low in cost and has ease of handling. The disadvantage of the plastic fiber is that it has high attenuation and less resistance than glass to higher temperatures. The major user of plastic fibers are the automobile, medical, and display sign industries. Automobile manufacturers use plastic fiber as sensors and display indicators for the car panel. Plastic fiber can transmit 60% of the incoming light over a 2 m distance. The medical industry has been using plastic fibers for at least 10 years. The fiber endoscopes that look at the tissues of the body have replaced large, inflexible scopes that doctors used previously.

3.14.2 Dispersion-Shifted Fiber

Dispersion-shifted fiber was invented to be used in the 1550 nm range. At 1550 nm, the waveguide and material dispersion cancel each other out. The best way to make the fiber is to change the waveguide dispersion. This is done by changing the refractive index profile to look something like a bull's-eye. The innermost ring is an inner core that has a triangular refractive index profile. This is surrounded by a lower-index silica region. The outer core surrounds this layer and has a step-index profile. The outer core is then surrounded by the cladding. Just from the description of the fiber's refractive index profile, this is a costly fiber to make. The fiber can be used to minimize the number of repeaters but must be handled very carefully to avoid microbending losses in the dispersion-shifted fiber.

3.15 SUMMARY

One of the distinguishing physical parts of a fiber optic cable is the size of the core compared to the cladding. The smallest fiber size, designated 9/125, is a single-mode fiber with a 9 μm core and a 125 μm cladding diameter.

Materials play an important role in fiber types. The three raw starting materials for fiber are fused quartz, silica dioxide for glass, and some type of plastic.

Launch conditions are defined as coupling the fiber to the source. The fiber must be able to transmit the optical power from a diffused source. The amount of light-gathering ability of the fiber is known as the numerical aperture.

The optical energy is classified into modes. For multimode fiber, hundreds of modes are propagated down the fiber; single-mode propagates only one. The ability to propagate these modes is based on the chemical makeup of the core–cladding interface. There are refractive index profiles that help guide the light down the fiber.

There are three main types of fiber: single-mode (for very long distance links), multimode graded-index (for moderate to long lengths), and multimode step-index fiber.

Dispersion influences the bandwidth, bit rate, and pulse shape of the fiber. Modal and material dispersion are found only in multimode fiber. Material and waveguide are found in single-mode. Bandwidth and bandwidth length product are used to measure the dispersion of the fiber. The bandwidth of an optical system is larger than the bandwidth of an electrical system.

Total attenuation of the fiber is due to scattering (imperfections in the glass) and absorption (electronic and atomic material properties). The most important attenuation loss is ion absorption, which causes the production of the fiber to be a clean environmentally monitored process. Bending losses (microscopic and macroscopic) occur during fabrication and installation of the fiber.

The cutoff wavelength is the lowest possible wavelength that a single-mode fiber can propagate without looking like a multimode fiber. Mode field diameter is more important than the core diameter because it more accurately describes the area in which the light propagates. Table 3-4 summarizes optical fiber specifications and their applications.

3.16 EQUATION SUMMARY

Far-field radiation pattern:

$$I = I_0 \cos \phi \qquad \text{watts/steradian} \qquad (3\text{-}1)$$

Total power Φ_0 emitted by the source:

$$\Phi_0 = \pi \cdot I_0 \qquad (3\text{-}2)$$

Amount of power that can be collected by the fiber (whose core diameter is greater than the source):

$$\Phi = \Phi_0 \cdot (NA)^2 \qquad (3\text{-}3)$$

TABLE 3-4 Optical Fiber Parameters and Applications

Fiber Type	Core/Clad Ratio (μm)	NA	Bandwidth (MHz-km)		Loss (dB/km)		Application
			At 850 nm	At 1300 nm	At 850 nm	At 1300 nm	
Step-index single-mode	9/125	0.12		2000	5.0	0.40	Very long distance
Glass/glass multimode	50/125	0.20	400	350–1300	3.5–3.6	2.0–7.0	Long-distance telecom and local networking
Glass/glass multimode	62.5/125	0.29	250	300–600	3.0–6.0	1.0–2.0	High-speed local networking and links
Graded-index multimode	100/140	0.26–0.29	20	800	3.5–6.0	1.2–2.0	High-speed local networking and links
Glass/plastic multimode	400/550	0.30	15		10		Low-speed links, high radiation environment
Step-index multimode	300/440	0.27	25		10		Low-speed links
Plastic/plastic multimode	1000/2000	0.47	20		150 at 660 nm		Very short links

Fraction of optical power from a diffused source that a fiber will propagate:

$$\frac{\Phi}{\Phi_0} = (NA)^2 \approx 2 \cdot n \cdot \Delta n \tag{3-4}$$

Normalized wave number:

$$V = 2\pi \cdot \frac{a}{\lambda} \cdot NA \tag{3-5}$$

Approximate number of modes in a step-index profile:

$$N = \frac{V^2}{2} \tag{3-6}$$

Number of modes for a graded-index fiber is

$$N = \frac{V^2}{4} \tag{3-7}$$

Additional length it takes the meridional ray to travel:

$$\Delta Z = \frac{Z \cdot (n_{core} - n_{cladding})}{n_{cladding}} \quad \text{meters} \tag{3-8}$$

Additional time it takes for the meridional ray to travel:

$$\Delta t = \frac{Z \cdot (n_{core} - n_{cladding}) \cdot n_{core}}{c} \quad \text{seconds} \tag{3-9}$$

Axial ray path:

$$z_t(min) = \frac{z}{\sin \phi(max)} = \frac{z}{\sin 90°} = z \quad \text{meters} \tag{3-10}$$

Meridional ray path:

$$z_t(max) = \frac{z}{\sin \phi(min)} = \frac{z}{\sin \phi_c} = z \cdot \frac{n_{core}}{n_{cladding}} \quad \text{meters} \tag{3-11}$$

Total path-length difference due to this modal dispersion:

$$\Delta_z = z_t(max) - z_t(min) = z \cdot \left(\frac{n_{core}}{n_{cladding}} - 1 \right) \quad \text{meters} \tag{3-12}$$

Refractive index difference:

$$\Delta = \frac{n_{core} - n_{cladding}}{n_{core}} \tag{3-13}$$

Difference in path length:

$$\Delta_z = z \cdot \frac{\Delta}{1 - \Delta} \quad \text{meters} \tag{3-14}$$

Total time delay due to modal dispersion:

$$\Delta_{t\,modal} = \frac{\Delta_z}{V_p} = \frac{n_{core} \cdot z}{c} \cdot \frac{\Delta}{1 - \Delta} \qquad ns \qquad (3\text{-}15)$$

Total time delay due to material dispersion:

$$\Delta_{t\,material} = D_m \cdot z \cdot \lambda_{3dB} \qquad seconds \qquad (3\text{-}16)$$

Time delay due to waveguide dispersion:

$$\Delta_{t\,waveguide} = \Delta_w \cdot \lambda_{3dB} \cdot z \qquad seconds \qquad (3\text{-}17)$$

Total dispersion:

$$\Delta_{t(total)} = \sqrt{\Delta_{modal}^2 + \Delta_{material}^2 + \Delta_{waveguide}^2} \qquad seconds \qquad (3\text{-}18)$$

Pulse width of the received light:

$$t_r = t_w + \Delta_{t(total)} \qquad seconds \qquad (3\text{-}19)$$

Theoretical maximum bit rate for a digital receiver:

$$B = \frac{1}{t_r} = \frac{1}{t_w + \Delta_{t(total)}} \qquad (3\text{-}20)$$

Estimation of the bit rate for a return-to-zero bit pattern:

$$B = \frac{0.441}{t_r} = \frac{0.441}{t_w + \Delta_{t(total)}} \qquad (3\text{-}21)$$

Electrical versus optical bandwidth:

$$BW_{electrical} = 0.707 BW_{optic} \qquad (3\text{-}22)$$

Cutoff wavelength:

$$\lambda_c = \pi \cdot \frac{2 \cdot a}{V_c} \cdot NA \qquad (3\text{-}23)$$

Mode field diameter:

$$w_0 = \frac{2.6}{V_c} \cdot a \qquad (3\text{-}24)$$

QUESTIONS

1. Name three types of dispersion.
2. If the fiber is 65/125, what is the size of the core and the cladding?
3. Name some of the refractive index profiles and define them.

4. What is the difference between bandwidth and bandwidth–length product?
5. Name and briefly describe the various types of attenuation.
6. What is scattering?
7. Which of the three types of absorption causes the most problems in the production process?
8. What is the difference between optical and electrical bandwidth?
9. Give an analogy to the cutoff wavelength.
10. What could have the most catastrophic effect, micro- or macrobending?
11. Why is the mode field diameter used in testing instead of the core size?

PROBLEMS

Use these specifications for the problems below.

TYPICAL DIMENSIONS FOR A MULTIMODE FIBER WITH
STEP-INDEX PROFILE:

Core diameter $2a$	100 μm
Cladding diameter	140 μm
Core refractive index n_1	1.48
Cladding refractive index n_2	1.46

TYPICAL DIMENSIONS FOR A SINGLE-MODE FIBER:

Mode field diameter $2w_0$	10 μm
Core diameter $2a$	9 μm
Cladding diameter D	125 μm
Core refractive index n_1	1.46
Cladding refractive index n_2	1.45562

1. Find the amount of power propagated by a glass fiber with NA of 0.24.
2. Find the V number for a multimode step-index fiber at 850 nm. Compare the number of modes to that in Example 3-2.
3. Find the refractive index difference Δ for a single-mode fiber.
4. Find the additional time it takes for the longest ray in a multimode step-index fiber of length 10 km and 35 km.
5. Calculate the path length of the meridional ray for a multimode fiber that is 25 km in length. What is the percentage difference from the axial ray?
6. Find the minimum value of total modal dispersion for a 30-km link.
7. Find the chromatic dispersion for Examples 3-5 and 3-6.

8. A LED system has a spectral width of 45 nm. Modal dispersion is 230.98 ns/km, material dispersion is 1.45 ns/km, and there is no waveguide dispersion. Find the bit rate of the receiver. Assume a 1-km link.

9. Find the cutoff wavelength for the multimode step-index fiber with a normalized frequency cutoff of 1.88.

10. Using the mode field diameter and the core diameter of the single-mode fiber, find the normalized frequency.

Fiber Fabrication
and Cable Design

The student will be able to:

- Become familiar with the various types of fabrication processes.
- Compare the merits of each fabrication process.
- Identify each part of a total cable configuration and its significance.
- Understand some of the testing performed on the cable before it is installed.

4.1 INTRODUCTION

The fiber fabrication process must be tightly controlled, both physically and environmentally. This requires that the optical fiber manufacturing process start with an ultrapure starting material: silica, glass, or plastic. Silica fibers get their name from the compound silicon dioxide (SiO_2). This compound is found in quartz or quartzite, one of the components of sand. Actually, a material called *fused silica glass* is used, which is a glassy, solidified melt of silica dioxide. Most optical waveguides are made up of a silica dioxide core and a metal oxide cladding. In general terms, the process used to make the fiber starts with pure silica dioxide with added dopants. Dopants are added to increase or decrease the refractive index. By adding oxides such as germanium dioxide (GeO$_2$) and phosphorus pentaoxide (P$_2$O$_5$), the refractive index is increased. To decrease the refractive index, fluorine (F) or boron trioxide (B$_2$O$_3$) can be added.

Glass cores and cladding materials are chosen so that they have similar viscosities, low melting temperatures, and long-term chemical stability. Also, the cladding must have a lower refractive index than the core. The next stage is to melt these glass materials into homogeneous, bubble-free glass rod. This glass rod, called the *preform* (Figure 4-1), is then taken to the drawing tower. The fiber is pulled into the diameters needed at the tower.

This discussion will focus on the production of glass fiber and the various

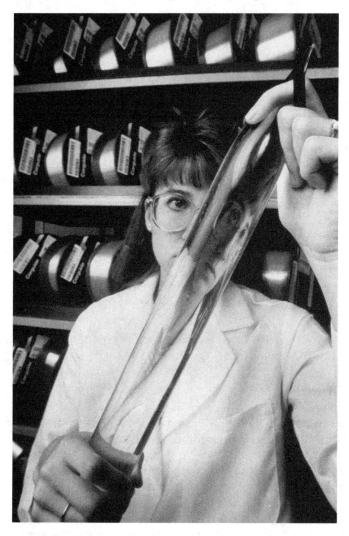

Figure 4-1 Preform: A process engineer at Corning's manufacturing facility inspects a glass blank made from ultrapure silicate glass before it is drawn into optical fiber. (Photo courtesy of Corning Incorporated.)

methods involved. Plastics are easier to manufacture and are chosen primarily for cost reasons. This process is not covered in this book.

4.2 FIBER FABRICATION

Six different types of fabrication techniques that have been popular are (1) the rod-in-tube method, (2) the double crucible method, (3) modified chemical vapor deposition (MCVD), (4) vapor-axial deposition (VAD), (5) plasma-activated chemical vapor deposition (PCVD), and (6) plasma-impulse chemical vapor deposition.

4.2.1 Rod-in-Tube Method

The rod-in-tube method was one of the first processes used to make cable. The core material of ultrapure fused silica glass rod is slid into a fused silica glass tube, or cladding with a lower refractive index. The rod and tube have only a minimal gap between them. This can be a disadvantage. Impurities, bubbles, and perturbations at that gap interface can cause attenuation losses, sometimes as high as 500 to 1000 dB/km. The combined structure is heated until the tube collapses, producing a preform. The preform is heated, causing the glass to soften so that it can be drawn into the optical fiber (Figure 4-2). This method could only be used for making multimode step-index fibers. A method called the double crucible method was developed to avoid the high-attenuation processes.

4.2.2 Double Crucible Method

The double crucible method uses purified molten glass in separate crucibles within a controlled furnace environment. The core crucible is inside the clad-

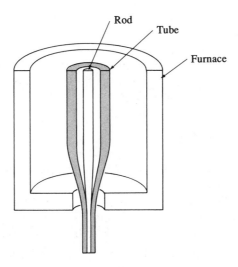

Figure 4-2 Rod-in-tube method.

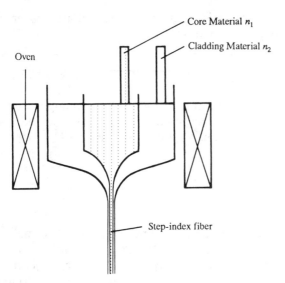

Core Material n_1

Cladding Material n_2

Oven

Step-index fiber

Figure 4-3 Double crucible method. (Courtesy of Les Editions Le Griffon D'Argile.)

ding material crucible. The crucibles are made of platinum so as not to interact with the glass material. They are shaped so that the bottom of each crucible narrows to a small nozzle, the diameter of which is the same as the desired core (inner crucible) or cladding (outer crucible). As the joined fiber leaves the crucible, it is drawn onto a takeup reel that is turning at a constant speed of 1 to 10 kilometers per hour (km/h). The environment that surrounds the crucibles is closely controlled to prevent contamination by oxygen and water vapor, which may cause bubbles in the glass. Some companies have a continuous feed of glass into the crucibles, thus achieving long, unbroken lengths of optical fiber (Figure 4-3). This is a cost-effective method for producing large-core-diameter fiber with a large numerical aperture.

One disadvantage of the double crucible method is that bubbles form in the melted glass. Also, the nozzle can cause the diameter of the fiber to fluctuate. Another disadvantage is that the fiber has a high water content, leading to degradation of the fiber after a relatively short period of time. Despite its disadvantages, the double crucible method produces very pure fiber. It is used to make graded-index fiber and step-index fiber. The attenuation of the fiber is 5 to 20 dB/km at 850 nm.

4.3 MASS PRODUCTION OF THE FIBER

In the previous methods discussed, the attenuation was very high. These methods, called *liquid phase deposition*, have attenuation losses between 5 and 20 dB/km. In 1970, Corning Glass Works used a technique called vapor deposition, which produced fiber that had an attenuation of 0.2 dB/km at 1550 nm. To make fiber in mass quantities, the glass fiber must be prepared in a two-stage process. The first process involves the making of a larger, exact replica of the fiber. The second stage involves pulling the rod to the dimensions of the fiber optic cable. The end product is a cable 20 to 50 km in length.

Example 4-1

A preform has a length of 2.0 m, an outer diameter of 1 cm, and a core diameter of 3 mm. From this preform, a fiber of outer diameter 125 μm is desired. Calculate the core diameter and length of the fiber that would be obtained.

SOLUTION

Outer-diameter reduction ratio:

$$\frac{1 \text{ cm}}{125 \ \mu\text{m}} = 80$$

Final core diameter:

$$\frac{3 \times 10^{-3}}{80} = 37.5 \ \mu\text{m}$$

Volume of preform:

$$V = \frac{\pi d^2 \cdot l}{4}$$

    V is the volume
    d is the outer diameter
    l is the length

$$\frac{\pi \cdot (1 \times 10^{-2})^2 \cdot (2)}{4} = 157 \times 10^{-6} \text{ m}^3$$

Assume that the volume of finished fiber will be the same as preform and solve for length (l):

$$l = \frac{4 \cdot V}{\pi d^2}$$

$$= \frac{4 \cdot (157 \times 10^{-6})}{\pi \cdot (1.25 \times 10^{-4})^2} \qquad (4\text{-}1)$$

$$= 1.28 \times 10^4 \text{ m} \quad \text{or} \quad 12.8 \text{ km}$$

4.3.1 Modified Chemical Vapor Deposition

The modified chemical vapor deposition uses a two-step process to produce the preform. A silica rod or bait rod, approximately 1 m in length with a 20 mm inner diameter, is rotated in a lathe or mandrel in the longitudinal direction. The outside is heated by an oxyhydrogen gas burner moving along the outside of the fiber in a very narrow region. The dopants needed for the refractive index, such as $SiCl_4$, $GeCl_4$, and PCl_3, flow through the inside of the tube. This produces the refractive index for the core material, while the outside is the material used for the cladding. To get a graded refractive index, 50 to 70 layers are deposited on the inside of the tube at 1600°C. This method occurs in an enclosed reactor, which means that the environment can be more closely controlled, thereby producing a very pure high-performance fiber.

In the second step, the rod is melted to 2000°C, shrinking the rod (Figure 4-4). The attenuation losses are roughly 3.4 dB/km at 0.825 μm and 1.2 dB/km at 1.3 μm.

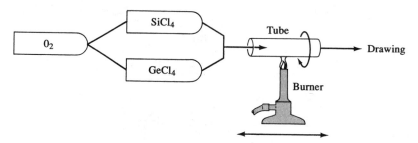

Figure 4-4   Modified chemical vapor deposition.

4.3.2 Outside Vapor Deposition

Outside vapor deposition (OVD) rotates a bait rod on the longitudinal axis in a turning lathe or mandrel. It is then heated on the outside by a propane burner. The necessary chemicals used to make the refractive index are fed into the burner and deposited as soot on the rotating rod. The rod is moved back and forth, building up the layers of the core and cladding. The process is stopped when enough layers have been deposited, then the bait rod is removed. The remaining soot rod is heated in segments over the entire length until it collapses. The melting point is between 1400 and 1600°C. This new rod is called a *blank*. Gaseous chlorine is used as a drying agent during this process to remove water particles in the air (Figure 4-5). The losses, using

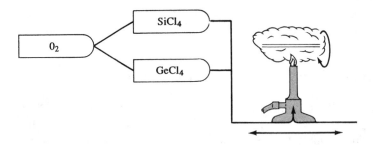

Figure 4-5   Outside vapor deposition.

this method, are about 1 to 1.8 dB/km at wavelengths of 1.2 and 1.55 μm. The disadvantage of this method is that the bait rod, when removed, can cause stress fractures on the inside wall of the soot rod. This is also a batch technique and does not produce long runs of the fiber.

4.3.3 Vapor-Axial Deposition

Vapor-axial deposition (VAD) is similar to OVD, but the soot particles are deposited using the end face of the rotating rod. The preform is drawn upward, keeping the distance between the burner and the preform constant. Several burners can be used at one time to create the different refractive index profiles. After the deposition process is completed, the preform is shrunk to a blank in a ring-shaped furnace. Gaseous chlorine is used to prevent the water content from building during this process. Attenuation losses range from 0.7 to 2.0 dB/km (Figure 4-6). The disadvantage of this method is that the OH^- impurity content cannot be controlled as desired because of the flame hydrolysis method.

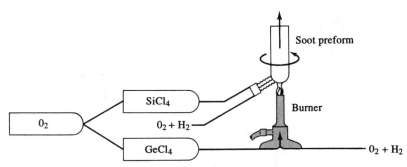

Figure 4-6   Vapor-axial deposition.

4.3.4 Plasma-Activated Chemical Vapor Deposition

Plasma-activated chemical vapor deposition (PCVD) is similar to the MCVD system, but instead of a flame, a plasma torch is used. Plasma is produced when a gas is excited by microwaves. The plasma consists of ionized electric charge carriers, which can be used to melt materials at high melting points. The reason this method is popular is that the flame can move quickly longitudinally across the rod, causing more layers (as many as 2000) to be produced. This allows for a more precise refractive index profile (Figure 4-7). The attenuation losses are between 3 and 4 dB/km at a wavelength of 0.85 μm.

4.3.5 Plasma-Impulse Chemical Vapor Deposition

This process has recently been developed. By using low-pressure discharge pulses, layers of glass are deposited simultaneously on the inner surface of the glass rod. This is advantageous because no moving parts are needed.

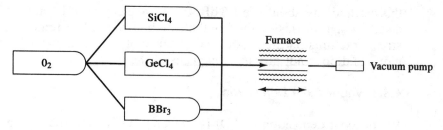

Figure 4-7 Plasma-activated chemical vapor deposition.

4.4 COMPARISON OF THE PROCESSES

Rod-in-tube method: moderate performance and low cost at moderate volume.

Double Crucible method: moderate performance and low cost at large volume.

MCVD: best performance but expensive.

OVD: good performance, less expensive.

VAD: moderate performance and low cost at large volume.

PCVD: moderate performance and moderate cost at moderate volume.

4.5 FIBER DRAWING PROCESS

To make the fiber in the core/cladding ratio necessary, the preform from any of the methods described previously, is taken to a drawing tower and attached to a mount. The mount is adjusted to a vertical feed mechanism, where the lower end of the preform is heated to 2000°C. To ensure that the diameters stay within specification, the drawing speed and the feed mechanism must be adjustable. The drawing speed is approximately 200 m/min. The diameter is monitored closely. The drawing process must be controlled to keep the dimensions (the diameter), strength, and the optical properties within specifications.

The circularity and concentricity of the core and cladding must be set when the preform is made. In the drawing process, the diameter can be affected if the preform feed rate and the drawing speed are not in proper balance. The strength of the fiber must be enhanced at this point. The fiber is very fragile and will degrade with age and environmental contamination. A coating is applied to improve fiber strength and simplify handling. More about the coatings later in the chapter.

To preserve the optical properties, the diameter and the strength of the fiber must be preserved. The drawing process must be controlled very closely. The volume of the material fed into the heat must be equivalent to the volume being removed. In most operations, the preform feed rate and the temperature of the heating mechanism are held constant while the drawing speed is varied to control the diameter of the fiber. The relationship for this is as follows:

$$S = s \cdot \frac{D^2}{d^2} \qquad (4\text{-}2)$$

S is the preform descent feed rate, in meters per second

s is the fiber drawing speed, in meters per second

D is the preform diameter, in meters

d is the fiber diameter, in meters

The fiber is then wound onto a takeup cylindrical reel and shipped off to become fiber optic cable.

Example 4-2

The preform is 260 cm. The fiber needs to have a 140 μm outer diameter. The fiber drawing speed is 200 m/min. What is the preform descent feed rate?

<u>SOLUTION</u>

$$S = 200/\text{min} \cdot \frac{(140 \times 10^{-6})^2}{(260 \times 10^{-2})^2}$$
$$= 5.79 \times 10^{-7} \text{ m/min}$$

4.6 COATINGS

The optical fiber is coated with a type of polymeric material. This is used to protect the fiber from abrasion, microbends, environmental/chemical attacks, and mechanical damage after the preform has been drawn (Figure 4-8). The fiber immediately after drawing is probably at its most pristine state. Anything coming into contact with the glass will cause abrasions on the glass surface; therefore, the addition of a coating will protect the fiber. The coating stiffens the fiber, protecting it from conforming to other surfaces close to it. Different-colored pigments and dyes are added to the coating. The various colors are added to the coatings to identify individual fibers in a multifiber bundle.

Coatings must be properly applied to the fiber so as not to squeeze the fiber, causing attenuation loss. Some methods that are used include dipcoating, extrusion, spray coating, and electrostatic coating. The coating dimensions, in all of these processes, must be held to a minimum. The fiber must

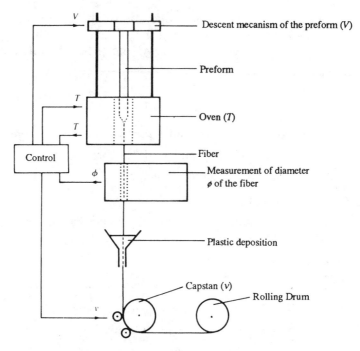

Figure 4-8 Coating process. (Courtesy of Les Editions Le Griffon D'Argile.)

also be well centered in the coating, so the uniformity of the coating is important, especially when splicing (joining) two fibers. A thinner coating gives better centering and better splicing.

Materials used for coating or sheathing depend on the particular environmental conditions. The most popular is polyethylene (PE), used for most outdoor and general use requirements. Polyvinyl chloride (PVC) is used for indoor cables where flame resistance is required. However, the PVC coating has been used for outdoor cables, where the cable plant is directly buried in soil. FEP (perfluorinated ethylene propylene) is used for applications where the outside temperature can reach 100°F or higher. This coating is highly resistant to weathering. Ethylene vinyl acetate (EVA) coating is used when

TABLE 4-1 Material System and Fiber Properties

Material System	Attenuation (dB/km)	Fabrication Method	NA	Fiber Core Diameter (μm)
SiO_2–P_2O_5	—	CVD	0.18	50
Silica–GeO_2	4	CVD	0.14	20
Soda-borosilicate	15	Double crucible	0.15	20
Alkali-lead silicate	30	Rod in tube	0.45	50
Soda-lime silicate	45	Rod in tube	0.20	30

external temperatures are higher than 250°F. This type of coating is sometimes known as flame-retardant noncorrosive (FRNC). Table 4-1 summarizes the starting materials and processes required to make the fiber.

4.7 CABLE DESIGN REQUIREMENTS

Before the optical fiber can be used in practical situations, the fiber must be protected from the environment. Fiber optic cable is installed in a variety of different environments, indoors and outdoors; therefore, the cable must be designed for different installation requirements. Three basic areas must be considered in cable design: optical, mechanical, and constructional.

The *optical* requirement for the cable design says that there must be enough optical fibers for the total required transmission capacity. Reserve or redundant fibers must also be added. The attenuation at the operational wavelength and the numerical aperture of each fiber must not change significantly during the installation or lifetime of the fiber. The same optical fiber or groups of fibers may be used in several different types of cable. For example, the same fiber may be used for aerial or underground installation. The difference would be in the outer cabling of the fiber.

The *mechanical* requirements for wire and optical cable include the maximum tensile load and the minimum bend radius. These requirements are critical in both the installation process and the lifetime of the cable. The mechanical requirement involves the tensile properties or the stress/strain behavior that must be met for a particular application. The minimum bend radius requirement for wire is determined by the construction and cable size. High-frequency coax cable (RG/8 or RG/11) has a minimum bend radius of 10 to 12.5 cm. The maximum tensile load for a coax cable (RG/58), the same size as a fiber optic cable, has a maximum load of 60 to 100 lbs. Compare this to a typical fiber optic cable with a minimum bend radius of 14.75 cm with a tensile force applied. With no force applied, the bend radius is 30 mm. The maximum force on a fiber optic cable should be limited to 90 lbs. An installer of fiber optic cable uses a tensiometer with a clutch, which will stop the pull process if the tensile strength is exceeded. The tensile strength is measured is newtons or pounds per square inch. The cable has a crush-resistance specification, which indicates the cable's resistance to radial compression. These forces are parallel to the surface. A typical parameter for crushing without mechanical damage is 57 lb/in². If the force is increased to 114 lb/in², deformation of the buffer tube will occur but will not affect the optical properties. This specification is measured in newtons/centimeters. The cable must also have some type of flexing resistance to allow for bending with a certain radius without causing microbends in the fiber. The outer jacket, which provides protection for the fiber, must have material that is stable. This means that the outer jacket will not corrode or degrade due to environmental conditions.

The *constructional* requirements are limited by the size of the optical fiber. The material and dimensions of the core, cladding, and coatings of the fiber must be considered before adding on more material to protect it. Remember that one of the main features of the fiber optic cable is its small size. The cable dimensions and the added protective coverings will have to be designed to meet the customer's installation requirements.

4.8 TYPICAL CABLE DESIGN

Figure 4-9 shows the parts of a typical cable design: fiber, coating, buffer tube, strength member, and outer jacket. There are many applications where a cable with a single optical fiber is sufficient to carry the required data traffic. There are also many applications where multiple fiber optic cables need to be used. Instead of installing multiple single cables, special-purpose cable is designed. A typical multiple fiber, loose buffer tube configuration is illustrated in Figure 4-9. Starting with the innermost element, a typical cable configuration will be discussed.

4.8.1 Central Member

The innermost element of the total cable is a support element called the central member. The central member facilitates stranding or braiding the fibers around a central point, preventing buckling and kinking. The central member is made of steel wire or a fiberglass/epoxy material. It is surrounded

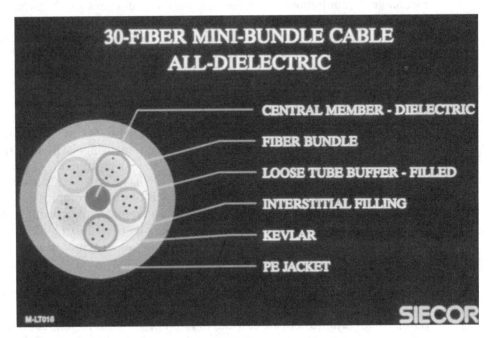

Figure 4-9 Total cable configuration. (Courtesy of Siecor Corporation, Hickory, North Carolina.)

by a buffer used as a cushioning layer. The individual fiber cables are stranded around the central member and consist of just the optical fiber, coating, and buffer tube.

4.8.2 Buffer Tubes

The main purpose of the buffer tube, is to enhance the tensile strength and provide radial protection for the fiber. Types of buffer tubes are tight, loose, and filled loose. The *tight buffer tube* (Figure 4-10) is usually a hard plastic and is in contact with the coated fiber. It is used primarily for outdoor applications, where it provides good protection for the fiber in most environmental surroundings. The *loose buffer tube* (Figure 4-11) allows the fiber to move inside it. This relieves the cable from the microbend and macrobend stresses that could be encountered. These stresses occur especially during installation or in situations were the cable requires frequent handling. The *filled loose buffer tube* is filled with a moisture-resistant compound between the fiber and the tube. This makes the cable more suitable for underground or underwater installations.

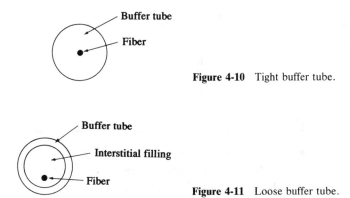

Figure 4-10 Tight buffer tube.

Figure 4-11 Loose buffer tube.

4.8.3 Stranding

There are two methods of stranding, helical and reverse lay (SZ), as shown in Figure 4-12. *Helical stranding* places the cable in one direction at a constant

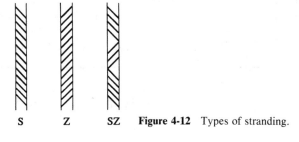

S   Z   SZ   **Figure 4-12** Types of stranding.

angle to the longitudinal axis of the fiber. *Reverse lay stranding* reverses the direction of stranding after a predetermined number of revolutions. To calculate how much excess cable would be needed, the following formulas are used:

$$L = S \cdot \sqrt{1 + \left(\frac{2\pi R}{S}\right)^2} \qquad (4\text{-}3)$$

$$\alpha = \tan^{-1}\left(\frac{S}{2\pi R}\right) \qquad (4\text{-}4)$$

L is the length of the stranding element, in millimeters

S is the lay length, in millimeters

R is the radius, in millimeters

$2\pi R$ is the circumference, in millimeters of the stranding circle

α is the angle of the stranding

The excess length Z due to stranding is calculated as a percentage:

$$Z = \frac{L - S}{S} \times 100\% = \left[\sqrt{1 + \left(\frac{2\pi R}{S}\right)^2} - 1\right] \times 100\%$$

$$= \left(\frac{1}{\sin\alpha} - 1\right) \times 100\% \qquad (4\text{-}5)$$

Example 4-3

> The production line is set up to use helical stranding. The lay length is 102 mm and the stranding radius 4.3 mm. Find the excess length due to stranding and the stranding angle.
>
> SOLUTION
>
> $$Z = \left[\sqrt{1 + \left(\frac{2\pi(4.3)}{102}\right)^2} - 1\right] \times 100\% = 3.4\%$$
>
> This means that for each 100 m of cable the stranding elements must be 3.4 m longer.
>
> $$\alpha = \tan^{-1}\left(\frac{102}{2\pi(4.3)}\right) = 75.16°$$
>
> The slope of stranding occurs at approximately 75°.

It is sufficient to use these equations for the helical and the reverse lay stranding.

4.8.4 Cable Core Filling

Surrounding the individual fiber cables is a filling compound known as *interstitial filling*. This is a chemically inert oil that neither freezes nor attacks the fiber cables. Filling compound is used primarily to make the cable watertight. If watertightness is not required, for example in indoor cable, the interstices of the cable core are wrapped with a thin layer or layers of thin plastic foil.

4.8.5 Strength Member

The strength member material is made out of Kevlar, Aramid, fiberglass, or steel wires. The main purpose of the strength member is to add tensile strength to the cable for installation purposes.

4.8.6 Outer Jacket

The outer jacket is made out of a type of polyester material, such as polyvinyl chloride, polyurethane, or polyethylene. Its main purpose is to protect the fiber and the inner layers of the cable from moisture.

4.8.7 Cable Sheath

A strength member, usually Kevlar, surrounds the entire configuration of fiber cables. The cable sheath provides an enclosure for the strength member, filling, fiber cables, and central member. The most popular cable sheath is the polyethylene (PE) sheath. The sheath preserves the fiber cable environment and protects the cable from the outside world, during both the installation and the lifetime of the fiber. This sheath must be scuff-, impact-, crush-, moisture-, chemical-, and oil-resistant. It must also have a wide temperature range and in some cases be flame retardant.

The manufacturer's identification thread is applied during sheathing. Siecor uses a red–red–green–black thread, and Siemens uses a green–white–red–white thread. A measuring tape can be added during this process which has a continuous meter marking printed on it.

4.8.8 Armoring

In some cases, armoring is added for specific reasons, such as protection against the rigors of installation. The armoring also protects against rodents, shields against lightning, and helps in the supporting of aerial cables. The armoring can be made of stainless steel, copper-clad stainless steel, or strands or braids of Kevlar. An additional final layer made of polyester material can surround the armoring.

4.9 TESTING

There are numerous optical, mechanical, and environmental tests that can be performed on single-mode or multimode fibers. In general, the most routinely performed tests are the mechanical and optical. In this section only a few of these tests are described.

4.9.1 Tensile Strength Test

This is an extremely important mechanical test for the designer of optical fiber systems. The designer of a fiber optic system uses the tensile-strength parameter of the cable to select whether the particular cable can be used in a particular application. For example, an aerial fiber must have a higher tensile strength than that of a conduit type. The manufacturer of the cable must ensure that the fiber has not lost its ability to carry the light even after it has undergone strong tensile forces.

When a cable is being pulled, stress concentrations occur along the cable which could lead to stress fatigue. Water intrudes into the stress fractures, causing fiber degradation or failure. A tensile strength test determines the elongation of the cable and the fiber as well as the attenuation of the fiber. During the test the cable is hooked up to an elongation tester. It is subjected to a constant tensile load for a period of time. The elongation of the fiber is determined by a pulse delay time measurement. The attenuation of the fiber is tested using four different wavelengths.

4.9.2 Other Mechanical Tests

Optical cables can be subjected to different stresses for their particular application. For instance, the fiber used by the U.S. Army must withstand a lateral crush (i.e., a tank rolling over the cable). For this reason there are tests for crush, impact, and vibration to determine these properties.

4.9.3 Environmental Test

This temperature test is important because with varied temperature changes the attenuation of the cable may change. This causes loss of transmission properties. These temperature changes can occur during storage and the lifetime of the cable. The cable is wrapped around a drum and placed in a computer-controlled temperature chamber. Both ends are attached to an attenuation measurement test set. Various temperature cycles are run on the cable and the results recorded. Cables must be durable and not disintegrate with time. Aging tests are carried out in high-temperature chambers over a half-year's time to record transmission property changes.

Other environmental tests are as follows:

Power transmission versus temperature: checks the effects of temperature on the transmission of the light through the fiber.

Power transmission versus humidity: checks the effects of humidity on the transmission of light through the fiber.

Power transmission versus ice crush: the fiber is frozen to see if the power transmission stays stable.

Flammability: the fiber is hung with a flame held under the fiber; toxic fumes and how far up the cable burns are factors in this test.

4.9.4 Optical Tests

Optical properties such as attenuation, cutoff wavelength, mode-field diameter, and far-field distribution are all standard tests for optical properties. Table 4-2 summarizes these properties.

Other measurements that are taken are:

Fiber size measurements: measures the outer diameter of the fiber.

Fiber bundle measurements: checks the bundle size.

Number of fibers: determines the number of fibers; the fiber cable is photographed and the number of fibers per bundle are counted.

Far-end crosstalk: two fiber specimens are placed parallel to each other; one is illuminated and the other is monitored for any light being received.

Refractive index profile: the fiber is tested using the near-field distribution to find the index profile.

Further tests can be performed at the request of the customer or by the particular company making the fiber.

TABLE 4-2 Typical Transmission and Optical Properties of Single-Mode and Multimode Fibers at Room Temperature

Property	Multimode Fiber 50/125		Single-Mode Fiber 10/125
	At 850 nm	At 1300 nm	at 1300 nm
Maximum attenuation coefficient (dB/km)	2.4–3.5	0.7–1.5	0.4–0.5
Minimum bandwidth for 1 km (MHz)	200–600	600–800	0–10 GHz
Maximum wavelength dispersion 1285 to 1330 nm (ps/μm)			3.5
Numerical aperture permissible tolerances	0.20 + 0.02	0.20 + 0.02	
Mode-field diameter permissible tolerances			10 + 1

4.10 SUMMARY

Various processes are used to make a fiber or a preform. Once the preform is made, it is pulled, coated, and put on a reel. The coated fiber then becomes a part of a total cable configuration. The cable may contain one fiber or hundreds of fibers. Additional supports or strength members are added to keep the cable rugged and strong for installation. While the cable is in the manufacturing facility, numerous tests are performed. Optical, environmental, and mechanical tests ensure that specifications are met before the cable can be installed.

4.11 EQUATION SUMMARY

Length of fiber from preform:

$$l = \frac{4 \cdot V}{\pi d^2} \qquad (4\text{-}1)$$

Preform descent speed:

$$S = s \cdot \frac{D^2}{d^2} \qquad (4\text{-}2)$$

Length of the stranding element:

$$L = S \cdot \sqrt{1 + \left(\frac{2\pi R}{S}\right)^2} \qquad (4\text{-}3)$$

Angle of stranding:

$$\alpha = \tan^{-1}\left(\frac{S}{2\pi R}\right) \qquad (4\text{-}4)$$

The excess length Z due to stranding:

$$Z = \frac{L - S}{S} \times 100\% = \left[\sqrt{1 + \left(\frac{2\pi R}{S}\right)^2} - 1\right] \times 100\%$$

$$= \left(\frac{1}{\sin \alpha} - 1\right) \times 100\% \qquad (4\text{-}5)$$

QUESTIONS

1. What is the raw material for glass fiber?
2. What is the least expensive method for making a fiber?
3. Why is the fiber coated?
4. What three requirements have to be met in cable design?

---

5. What is the difference between tight buffer and loose tube?

6. What is the significance of the strength member of the fiber? Where is it located?

7. What is the purpose of the central member?

8. What is the purpose of armoring?

9. What kind of coating could be used for an indoor cable that also must meet stringent fire codes for New York City?

10. If a college were to decide to manufacturer cable and wanted to maintain the cost, performance, and the volume at a moderate level, what process would you choose?

PROBLEMS

1. Using the double crucible method, the takeup reel has a constant speed of 5 km/h. How much fiber will be on that reel in 2 days if the diameter of the fiber is 125 μm?

2. What is the ratio difference if the preform is 20 mm in diameter and ends up as 50 μm?

3. If the lay length S is equal to 103 mm and the stranding radius R is equal to 4.5 mm, find the length of the stranding element L.

4. Using the lay length and stranding radius in Problem 3, find the stranding angle and the excess length Z. For a 4 km cable, how much longer should it be?

5. Using the formula below, find the elongation of a 16-fiber loose tube buffer cable. ΔR is defined as the difference between the outside diameter of the fiber and the inside diameter of the buffer tube. The cable's outside diameter (R) is 15 mm and the inside diameter of the buffer tube is 1.0 mm. Use the same lay length and stranding radius as in Problem 3.

$$\varepsilon = -1 + \sqrt{1 + \frac{4\pi^2 R^2}{S^2} \cdot \left(\frac{2\Delta R}{R} \pm \frac{\Delta R^2}{R^2} \right)}$$

minus sign is for elongation

plus sign is for contraction

6. Find the cable contraction for Problem 5.

5

Connectors, Splices,
and Couplers

The student will be able to:

- Describe the losses associated with joining fibers together.
- Know the process to prepare the fiber for splicing or connectorizing.
- Be able to identify various splices and connectors.
- Know and understand various coupler types.

5.1 INTRODUCTION

Connecting fiber optic cables is a critical part of the system. No matter what type of joining technique is used, the ultimate goal is to let the light go from one point to another, with as little loss as possible. Splicing unites two fibers into one continuous length. Connectors are used to couple from a transmitter to the fiber. Couplers are used to split information in many directions. Splicing, connectors, and couplers must be used and installed with low loss, high tolerances, and economical techniques. This chapter covers the various techniques used in the joining of fibers by using splices, connectors, couplers, or switches.

5.2 FIBER-TO-FIBER CONSIDERATIONS

Fibers have physical variations that cannot be eliminated during the manufacturing process. These variations vary from fiber to fiber as well as along the entire length of the fiber. There are two fiber losses that can occur: intrinsic and extrinsic.

5.3 INTRINSIC LOSSES

This type of loss occurs when the fiber itself is faulty, usually caused during manufacturing. Intrinsic losses are illustrated in Figure 5-1.

5.3.1 Core Mismatch

One type of intrinsic loss is caused by core area mismatch, which occurs when the fiber core of one cable is larger or smaller than the other. Typically, the diameter tolerance of a fiber is ±5% for both the core and the cladding. For two identical fibers with a ±5% tolerance, the loss from the diameter mismatch could be 0.83 dB. The formula for calculating the loss is

$$\text{loss}_{\text{dia}} = -10 \log \left(\frac{\text{dia}_r}{\text{dia}_t} \right)^2 \tag{5-1}$$

loss_{dia} is the loss due to diameter mismatch, in decibels

dia_r is the diameter of the receiving fiber

dia_t is the diameter of the transmitting fiber

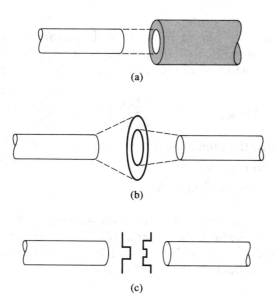

(a)

(b)

(c)

Figure 5-1 Intrinsic losses: (a) core area mismatch; (b) numerical aperture mismatch; (c) profile mismatch.

Example 5-1

Suppose that two fibers are to be joined. The transmitting fiber has a core diameter of 9 μm and the core diameter of the receiving fiber is 100 μm. What would the loss be?

SOLUTION

$$\text{loss}_{\text{dia}} = -10 \log \left(\frac{100 \ \mu\text{m}}{9 \ \mu\text{m}} \right)^2 = -20.92 \ \text{dB}$$

There will be no loss, since the light is passing from a smaller core to a larger core.

What if the situation were reversed, that is, the transmitting fiber is 100 μm and the receiving is 9 μm?

$$\text{loss}_{\text{dia}} = -10 \log \left(\frac{9 \ \mu\text{m}}{100 \ \mu\text{m}} \right)^2 = +20.92 \ \text{dB}$$

+20.92 dB is a lot of loss and is not acceptable.

More realistically, what if the diameters of the cores were within ±5%. What would the loss be? Using a 95 μm receiving and a 100 μm-core-diameter transmitting fiber, what would the loss be?

$$\text{loss}_{\text{dia}} = -10 \log \left(\frac{95 \ \mu\text{m}}{100 \ \mu\text{m}} \right)^2 = +0.446 \ \text{dB}$$

This is a small loss and is acceptable.

5.3.2 Numerical Aperture Mismatch

Other intrinsic losses may be caused by numerical aperture mismatch or profile mismatch. In the case of *numerical aperture mismatch*, the NA of the transmitting fiber is larger than that of the receiving fiber.

$$\text{loss}_{\text{NA}} = -10 \log \left(\frac{\text{NA}_r}{\text{NA}_t} \right)^2 \qquad (5\text{-}2)$$

loss$_{\text{NA}}$ is the loss due the numerical aperture in dB

NA$_r$ is the NA of the receiving fiber

NA$_t$ is the NA of the transmitting fiber

The actual NA of the transmitting fiber varies with the source, fiber length, and the modal patterns. Therefore, the material NA should not be used, but rather, the output NA of the fiber. No NA mismatch occurs when the receiving fiber has an NA greater than the transmitting fiber.

Example 5-2

The NA of the transmitting fiber is 0.34 and the NA of the receiving fiber is 0.32. What is the loss?

SOLUTION

$$\text{loss}_{NA} = -10 \log \left(\frac{0.32}{0.34} \right)^2 = +0.526 \text{ dB}$$

5.3.3 Profile Mismatch

In *profile mismatch* the geometry between the core and the cladding of the two fibers to be joined does not coincide. The profile of the transmitting fiber may be more elliptical, and when joined with a more circular fiber, would result in a minimum region where the light could be transmitted.

5.4 EXTRINSIC LOSSES

Extrinsic losses are usually the result of operator error when joining the fiber. These losses are caused by lateral offset, angular misalignment, and end separation, as shown in Figure 5-2.

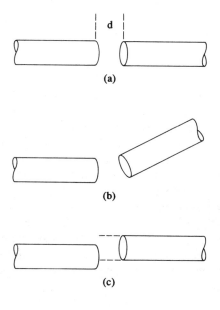

Figure 5-2 Extrinsic losses: (a) end separation; (b) angular misalignment; (c) lateral offset.

5.4.1 End Separation

When fibers are brought together, theoretically the fibers should meet. In most splices they do, but in connectors they are held slightly apart to prevent rubbing or damaging their end finishes. Gap loss depends on the width of the gap and the NA values of the fibers. From previous chapters, light spreads out in a conical shape. This defines the angle of the output cone of the transmitting fiber, more commonly known as the NA of the fiber. The width of the gap defines the portion of light that can be propagated by the receiving fiber.

Fresnel reflection loss is due to the difference in the refractive indices of the two fibers and the material separating them. A portion of the light is reflected back and is lost. These losses can be overcome by using an index-matching fluid to surround the fiber end faces. This is called a *wet joint*.

5.4.2 Angular Misalignment

Angular misalignment is not as great a loss as lateral displacement because the connector or splice can align the fiber more easily. The NA of the fiber affects the angular misalignment.

5.4.3 Lateral or Axial Displacement

Lateral or *axial displacement* is the result of two fibers not aligning correctly on their center axes. The lateral displacement causes the greatest loss in a splice or interconnection. This loss occurs even if there are no intrinsic variations in the fiber. A loss of 0.5 dB means that there is a displacement of about 10%. Connector manufacturers try to limit the lateral offset to 5% of the core diameter.

5.5 PREPARATION OF THE FIBER

The fiber end needs to be perpendicular and must have a mirror-like finish for any fiber connection or splice. The end face should be between 1 and 2° perpendicular. Fractures, burrs, surface roughness, or defects change the propagation of the light and cause scattering. The cable must be stripped to prepare the fibers for splicing or connectorization. The length in which the individual layers have to be stripped is dependent on the cable configuration. The outer jacket of the total cable configuration can be removed by a rip cord (a Kevlar string) being pulled up. The armoring is removed by another rip cord or by pliers, peeling back the armoring at the seam. Metal shears are used to cut the Kevlar or strength member because regular scissors would become dull after one cutting. The buffer tubes are stripped using electrical wire strippers. The plastic coatings can be removed by using chemicals such as acetone, or mechanically by using specially made wire strippers. Finally, the fiber is exposed, but it is still not perpendicular or free of blemishes. Two common methods used to prepare the fiber are the scribe and break method and the polishing method.

The *scribe and break method* is used primarily for splices. The cutting tool with a sharp blade made from diamond or tungsten carbide comes in various forms. The cutting tool may be a retractable blade in a pen casing or a cleaving tool, both shown in Figure 5-3. To use the pen cutting tool, hold the fiber on a work surface and scrape the pen cutting tool across the fiber, making a scratch. Take the free end of the fiber and bend until the fiber is broken. If done properly, the fiber will have a perpendicular end face. The cleaving tool aligns the fiber in the tool. The blade comes down on the cladding, making a small nick. After the nick has been made, pressure is released in the mechanism and the fiber breaks.

The *polishing method* is used as the last step after the connector is put on. The connector fits in a polishing fixture to ensure perpendicularity. The polishing fixture is moved across fine sandpaper in a figure-eight motion,

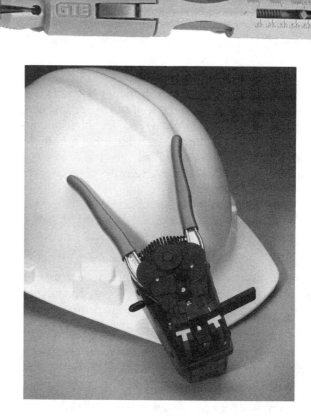

Figure 5-3 Cutting tools. (Courtesy of GTE.)

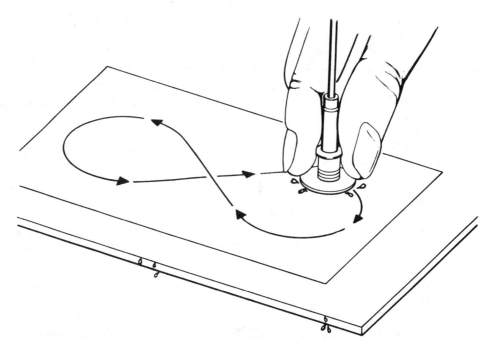

Figure 5-4 Polishing method. (Courtesy of Amp, Inc.)

shown in Figure 5-4. The fine sandpaper has grits from 1 to 0.3 μm. The fiber should be cleaned with reagent-grade isopropyl alcohol, a refined alcohol, each time the sandpaper is changed. The fiber and connector face will have a mirror-like finish after these steps have been completed.

No matter which method is used, the fiber should be inspected under a microscope. Figure 5-5 shows examples of acceptable and unacceptable polishes.

5.6 SPLICES

Splicing the fiber differs from using connectors. A splice welds, glues, or bonds together two ends of a fiber. This is considered a more permanent joint than that created by a connector. Splices are used for long-haul, high-capacity systems, while connectors are used for short-distance and end terminal equipment. Losses are approximately 0.2 dB. The three primary methods of splicing a fiber are the mechanical splice, the fusion splice, and the GTE Fastomeric splice.[1]

---

[1] Fastomeric splice is a trademark of GTE Fiber Optic Products, Williamsport, PA.

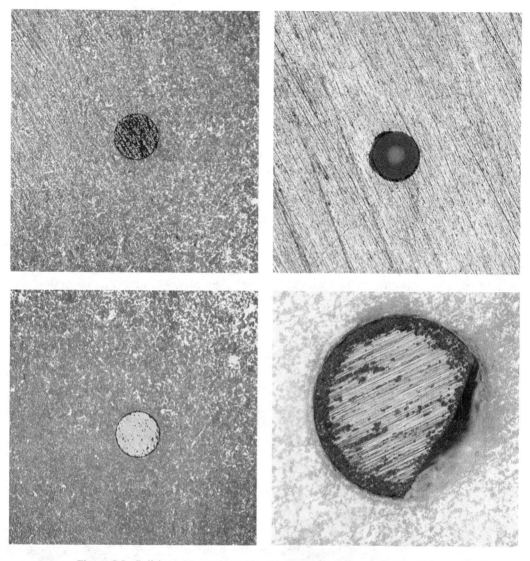

Figure 5-5 Polishes: (a) unacceptable; (b) acceptable; (c) acceptable; (d) unacceptable. (Courtesy of Buehler Limited, Lake Bluff, Illinois.)

5.6.1 Mechanical Splice

The mechanical splice is used mostly in the field. The fibers are self-aligned in a V-groove splice connector. The fiber ends are pressed together and are joined permanently using a rapidly curing index-matching epoxy. The epoxy is cured thermally or by ultraviolet light. The splice connector is then crimped onto the buffer tubes, which are held in position by pivoting arms on the mechanical splicer to provide stress relief. The splice losses that can be achieved by this splicing method range from 0.11 to 0.13 dB.

5.6.2 Fusion Splice

The fibers must be positioned with extreme care when using a fusion splicer. The splicer contains its own fiber cutting tool that has a carbide tip for clean mirror-like fractures. The fibers must be aligned on the X-Y-Z axes. Some fusion splicers have a X-Y-Z positioner; while others use ceramic V-blocks that align the X-Y axes automatically. This is to simplify adjustment in the Z axis. Using a microscope built into the splicer, the fibers are aligned. When the fibers are aligned in the proper position, the operator pushes a button that causes an electric arc to appear across the fiber ends. The arc melts the fibers together. Fusion splicing is not as difficult as the description indicates because most fusion splicers have an automated alignment procedure. The operator looks at the ends of the fiber and if everything is aligned properly, presses a button. The joined fibers are then protected by a metal splice connector that is crimped over the buffer tubes. The fusion splice loss is from 0.05 to 0.07 dB. A fusion splicer is depicted in Figure 5-6.

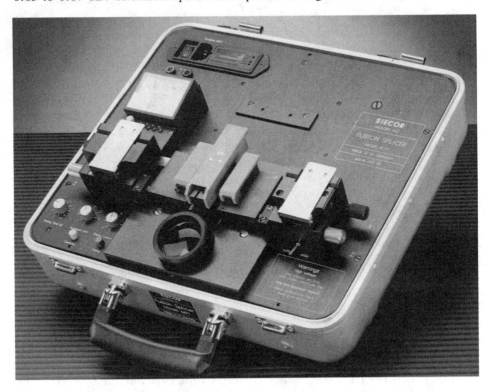

Figure 5-6 Fusion splicer. (Courtesy of Siecor Corporation, Hickory, North Carolina.)

5.6.3 GTE Fastomeric Splice

Another popular device for splicing is the GTE Fastomeric splice sleeve. The fusion splice method requires training for the technician performing the splice. The GTE Fastomeric splice, on the other hand, is easy to use. The fibers are cleaned and cleaved, as if preparing for a fusion splice. They are then inserted into a precisely formed V-groove inside the elastomer sleeve (Figure 5-7). An index-matching fluid acts as a lubricant for the fiber, preventing reflective losses in the fiber. The top lid of the GTE Fastomeric housing is then snapped into place. The splice assembly is ready to be installed in the splice case. Splice loss using the GTE Fastomeric splice is usually under 1.0 dB.

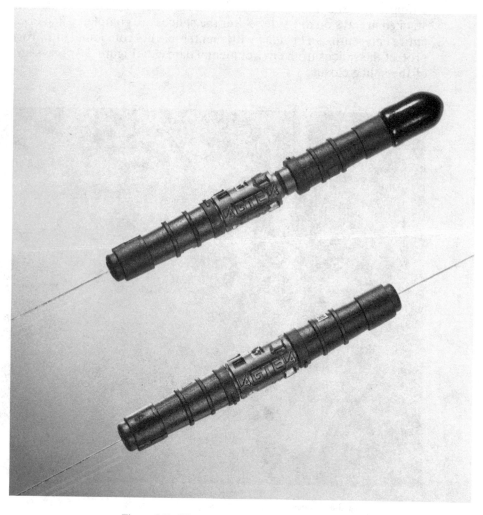

Figure 5-7 Elastomeric splice. (Courtesy of GTE.)

5.6.4 Protecting the Splice

Fusion, mechanical, and GTE Fastomeric splicing techniques have been discussed so far. Simple splicing is often only a small part of fiber optic field installation. In some installations, over 2000 splices will have to be made. If the cables are not neatly arranged in some type of closure there is a possibility of fiber breakage later. The closure encloses the splices, totally encasing the splices and a short length of cable on each side of the splice. Splices must be protected from such environmental conditions as water, aerial applications (wind), burial (dirt), and rodent protection (plow-in and direct burial). Most splice closures are similar to those used for copper cables, commonly called *UC closures*. The body is made of polypropylene copolymer with an inner metallic frame for mechanical rigidity and electrical shielding. The cable is wrapped around in a figure eight inside. The splices are then made and stacked in trays on one side. After the splices are completed and checked, the entire enclosure can be filled with reenterable gel compound or nitrogen gas to protect the splices from environmental damage. Figure 5-8 shows an example of the splice closure.

Figure 5-8 Splice closure. (Courtesy of Siecor Corporation, Hickory, North Carolina.)

5.7 CONNECTORS

Connectors are sometimes called nonpermanent joints. They are used to connect optical fibers to transmitters and receivers or panels and mounts. Connectors are increasingly easier to handle, mount, and install. Specific directions are used to prepare the fiber for the particular connector. The type of epoxy or cementing agent that must be used, the length of the jacket, the strength member, and the fiber that must be stripped back are specified in the directions. To put a connector on takes about 5 minutes, with up to 10 minutes for curing of the epoxy.

There are many different types of fiber optic connectors. Most can be used for the same type of fiber optic cable. The choice is usually up to the designer of the system or is based on the optical transmitters and receivers to

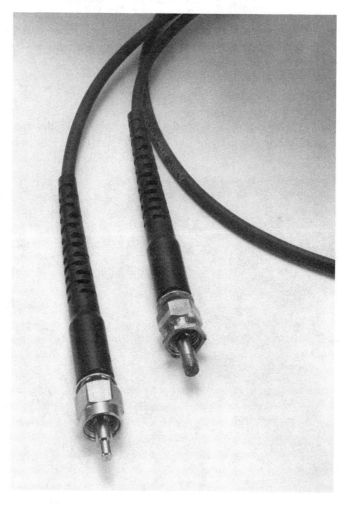

Figure 5-9 SMA connector. (Courtesy of AMP, Inc.)

be used. Because of its small core diameter, single-mode fiber is harder to connectorize and splice than is multimode fiber. Connectorizing single-mode fiber can be quite costly because high-precision connectors are required.

Two major areas in which connectors have wide usage are the telecommunications and data communications fields. In telecommunications, biconic, STC, FC, PC, and D3/D4 are used; in data communications the biconic, SMA, STC, and D4 are used. These types of connectors are discussed in the following sections.

5.7.1 SMA Connector

The SMA connector was designed from a similar connector used in the military or for microwaves. A retention clip surrounds the fiber and is pressed into the plug. The receptacle contains the other fiber, also held by a retention clip. When mated in a straight precision sleeve, the fibers are forced to be centered. This connector can maintain axial alignment to within 0.10°. The SMA is made out of porcelain or aluminum and is shown in Figure 5-9. Losses are on the order of 1.0 to 1.5 dB.

5.7.2 STC Connector

The STC connector (Figure 5-10) was originally developed by AT&T. The STC connector is very similar to the electrical BNC connector, which has a

Figure 5-10   STC connector. (Courtesy of OFTI.)

keyed design. These types of connectors are generally used in local area networks. Losses are typically in the range of 0.5 dB.

5.7.3 Biconic Connector

A biconic connector is a silica-loaded epoxy resin plug that has a taper on one end (Figure 5-11). The taper mates to a free-floating alignment sleeve within the adapter. The silica–epoxy resin expands and contracts with the glass fiber, holding it in place inside the connector. Losses for this connector are approximately 0.5 dB. The biconic connector is used for telephony applications.

5.7.4 FC and PC Connectors

The FC (face-end) and PC (point contact) connectors use a compression collar to hold the strength members from the cable against the connector. A threaded nut fastens over the strength members. The fiber rests in the ferrule. The FC connector has a flat ferrule end and the PC connector has a pointed end. The fiber fits in a spring-loaded ferrule made of alumina ceramic for the FC and all ceramic for the PC. The alignment is extremely precise for these connectors but will not accommodate a cable larger than 3 mm in diame-

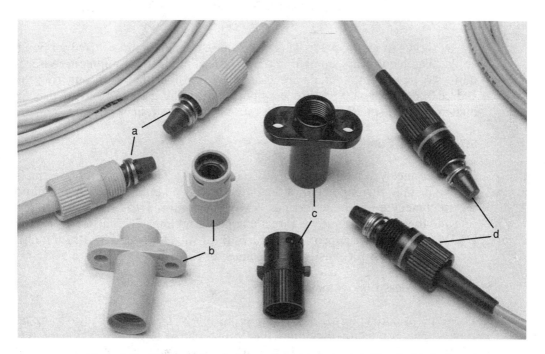

Figure 5-11 Biconic and conical connectors: (a) single-mode biconic; (b) single-mode couplings (flanged and built out); (c) multimode couplings; (d) multimode biconic. (Courtesy of 3M Corporation.)

Figure 5-12 FC and PC connector. (Courtesy of OFTI.)

ter, as shown in Figure 5-12. This limits these connectors to be used for patch cords. Patch cords are small fiber optic cables linking test equipment to transmitters or receivers. Pigtails are short pieces of fiber that stick out from the source or detector. Losses are less than 0.5 dB. These connectors are used in high-speed transmission systems because of the low reflections.

5.7.5 D3/D4 Connector

D3/D4 connectors mate with the FC connectors. This type of connector is easy to install. Strip the fiber of the outer jacket and buffer and cement it into the steel-jacketed ceramic ferrule. A rubber strain relief is slid over the ferrule and the entire assembly is threaded into a coupling nut. The losses for the D3/D4 connector are approximately 0.7 dB for single-mode, 0.35 dB for multimode, and 1.0 dB for plastic. The connectors are shown in Figure 5-13.

5.7.6 Epoxyless Connector

This type of connector is low cost and has a low loss (0.47 dB) for using no epoxy. The connector combines the spring-loaded ferrule from the FC/PC connector with the cable retention from the SMA and the locking mechanism of the STC connector. The XTC connector, as it is known, was designed by OFTI specifically for use in the field. The connector is shown in Figure 5-14.

Figure 5-13 D3/D4 connector. (Courtesy of OFTI.)

Figure 5-14 Epoxyless connector. (Courtesy of OFTI.)

5.8 COUPLERS

Fiber optic couplers carry optical signals from one or more entrances to one or more exits. The couplers can be used for single-mode or multimode fibers. Their biggest application is for local area networks (LANs). For instance, a simple LAN may be a computer linked to many engineering workstations. Information must be sent to other terminals or computers at the request of the user. Couplers must be added if the LAN uses fiber optic cables. They can be used for multiplexing, isolation, or to provide bidirectionality for the system.

When using wavelength-division multiplexing, one fiber can carry more than one signal simultaneously using different wavelengths. A single fiber, using a bidirectional coupler, may be used to both send and receive optical signals. Couplers may also be used to divide an optical signal from a single fiber across multiple fibers. For instance, a three-port coupler splits the incoming signal into two outgoing channels.

5.8.1 Passive Couplers

Passive couplers divide the signal into two or more parts. Two classic coupler types are the star and the tee coupler.

5.8.2 Star Couplers

The star coupler evenly distributes the incoming light from any one optical fiber or port to maximum of 64 fibers or ports. The star coupler is the central distribution point in a data bus network. There is a limit to the number of ports to which this coupler can divide the light. The two principal types of star couplers are the transmissive and the reflective. *Transmissive star couplers* are directional and provide isolation of the signal from other ports. A *reflective star coupler* distributes any input power to all ports, including its own, which means that this type of coupler can be tapped.

5.8.3 Tee Couplers

The tee coupler has three ports: one incoming, one outgoing, and the third going to a tap-off. The input light does not have to be equally divided among the three fibers but can be made with different power-splitting ratios. Tee couplers can provide a tap into a data bus at a selected point. The T-coupler system is limited by the number of taps, which reduces the signal level below receiver requirements (Figure 5-15).

5.8.4 Coupler Loss

Coupler loss can be excessive in a designer's loss margins and must be calculated. The total output power can be no more than the input power. This loss

---

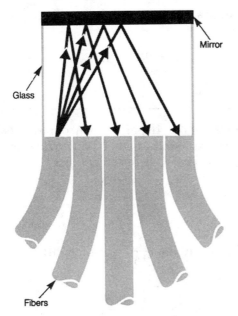

Glass

Mirror

Fibers

Figure 5-15 Star and tee couplers. (Courtesy of Cooper Industries, Inc., Belden Division.)

in the coupler is the ratio of the output power divided by the input power in decibels. If the output power is equally divided, the loss is at least 3 dB. Additional loss above the theoretical limit is called the excess loss. Three types of losses are usually calculated: insertion loss (IL), splitting loss (SL), and excess loss (EL).

Insertion loss is a measure of power lost by a signal propagating from a specific input port to a specific output port. To calculate the insertion loss, the following formula is used:

$$IL = -10 \log \left(\frac{P_{output}}{P_{input}} \right) \qquad (5\text{-}3)$$

IL is the insertion loss, in decibels

P_{output} is power output

P_{input} is power input

Splitting loss is the division of power among the coupler output ports and is calculated by using the equation

$$SL = -10 \log \left(\frac{1}{N} \right) \qquad (5\text{-}4)$$

SL is the splitting loss, in decibels

N is the number of output ports

Excess loss is the difference between the measured insertion loss and the splitting loss:

$$EL = IL - SL \qquad (5\text{-}5)$$

Example 5-3

A coupler splits its power among 16 ports: 1 mW of power is injected into a port and 10 μW is measured at another port. What are the splitting loss, insertion loss, and excess loss of the coupler?

SOLUTION

Splitting Loss:

$$SL = -10 \log \left(\frac{1}{16}\right) = +12.04 \text{ dB}$$

Insertion loss:

$$IL = -10 \log \left(\frac{1 \times 10^{-6}}{1 \times 10^{-3}}\right) = +30 \text{ dB}$$

Excess loss:

$$EL = 30 - 12.04 = 17.96 \text{ dB}$$

5.8.5 Coupler Design

There are three popular types of couplers: the fused biconical tapered coupler, wavelength-selective coupler, and active coupler. The *fused biconical tapered coupler* produces low crosstalk. It is made by bringing two bare fibers together, twisting them one and a half turns. The twist is melted and the region is carefully pulled apart. This causes the claddings to fuse together. The loss through a fused biconical coupler is about 1 dB. A diagram of the biconical coupler is shown in Figure 5-16.

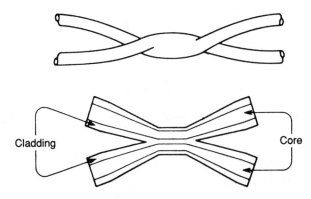

Cladding

Core

Figure 5-16 Biconic fused coupler. (Courtesy of Delmar Publishers Inc., Albany, New York: Sterling, *Technician's Guide to Fiber Optics.*)

Wavelength-division multiplexing is used exclusively by fiber optic signals. Signals from various sources transmit at different wavelengths. These signals are combined and sent over one fiber cable. The component that makes this possible is a *wavelength-selective multiport coupler*. Figure 5-17 shows this coupler. A diffraction grating is used to spread out the signal, directing it to the output fibers. This can be used at the input fiber.

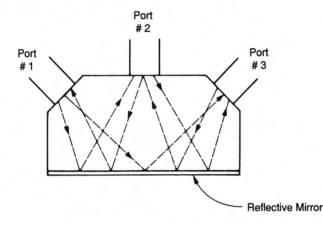

Figure 5-17 Wavelength-selective coupler.

Active couplers consist of receivers mated with transmitters. These couplers can detect the optical signal, regenerate it, and transmit it to other fibers. This is a type of repeater but uses pin detectors and light emitting diodes, discussed in subsequent chapters.

5.9 SWITCHES

Fiber optic switches can reroute the optical signals in a distribution system, or in loopback testing for faults or breaks in the cable. These switches are not easily implemented. A two-position switch consists of a GRIN lens and a sliding prism, as depicted in Figure 5-18. At fibers 1 and 2, light is being coupled into the GRIN lens. Looking at the output of fiber 1 only, the GRIN lens collimates the diverging beam emitted by the fiber. The right-angle prism

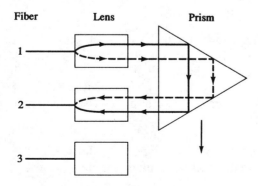

Figure 5-18 Two-position switch.

using total internal reflection of its slanted sides deflects the light and the GRIN lens focuses the beam into fiber 2.

A bypass switch is shown in Figure 5-19. In Figure 5-19, ports 1 and 4 are coupled and the other ports are isolated from them. When fiber 2 is pushed into fiber 1, ports 1 and 2 are coupled. A bypass switch is usually used in conjunction with a tee network in which the network can be included or not.

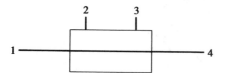

Figure 5-19 Bypass switch.

There are several concerns that must be evaluated when using a switch. The insertion loss of each port is

$$L_{il} = -10 \log \left(\frac{P_2}{P_1}\right) \tag{5-6}$$

L_{il} is the insertion loss, in decibels

P_2 is power output at port 2

P_1 is power input at port 1

Insertion loss depends on fiber alignment. Losses less than 1.5 dB can be obtained. Crosstalk is a measure of how well the uncoupled port is isolated. It is calculated by using formula (5-6). Typical values for this loss are 40 to 60 dB.

Switching speed is defined as how fast the switch can change from one position to the other. Switching is usually done electromagnetically. A magnet attracts a magnetic material to which an optical device (lenses, mirrors, prisms, and fibers) can be attached. When the electromagnet is turned off, a spring pulls the magnet back to a resting position. Switching times are on the order of a few milliseconds.

5.10 SUMMARY

Joining fibers together is not an easy task. Intrinsic losses such as core misalignment, numerical aperture mismatch, and profile mismatch may not be obvious when mating fibers. These problems are built into the fiber, although the manufacturing process is really to blame. Different types of extrinsic losses, such as gap loss, angular loss, and axial displacement have to be eliminated inside the connector or splice.

Before the fiber is put into the splice or connector, it is critical that it have a smooth flat perpendicular face. Specially designed tools are used to perform the scribe and break method, which can ensure a perpendicular end face. Now the fiber can be put into the splice. If a connector is being used, the fiber still has to be inserted into the connector and epoxied. The end of the connector is then polished to a mirror-like finish.

Couplers are used to direct the signal to one or many fibers. Different types of couplers can be used. Most common are the star and tee coupler. The star coupler routes the optical signal from one to many fibers. The tee coupler is a three-port device that taps off a portion of the signal to the third port. Coupler loss must be considered when designing a system.

Fiber optic switches are used to reroute an optical signal. Most of these switches use special lenses or prisms to accomplish the task. Different splice types, connectors, couplers, and switches are used to route the optical light to other fibers or to other transmitters and receivers. In later chapters, all of these components will be used with transmitters and receivers to form a complete fiber optic system.

5.11 EQUATION SUMMARY

Core mismatch:

$$\text{loss}_{\text{dia}} = -10 \log \left(\frac{\text{dia}_r}{\text{dia}_t} \right)^2 \tag{5-1}$$

Numerical aperture loss:

$$\text{loss}_{\text{NA}} = -10 \log \left(\frac{\text{NA}_r}{\text{NA}_t} \right)^2 \tag{5-2}$$

Insertion loss:

$$\text{IL} = -10 \log \left(\frac{P_{\text{output}}}{P_{\text{input}}} \right) \tag{5-3}$$

Splitting loss:

$$\text{SL} = -10 \log \left(\frac{1}{N} \right) \tag{5-4}$$

Excess loss:

$$\text{EL} = \text{IL} - \text{SL} \tag{5-5}$$

Insertion loss for a switch:

$$L_{\text{il}} = -10 \log \left(\frac{P_2}{P_1} \right) \tag{5-6}$$

QUESTIONS

1. Name the two types of couplers and describe them.
2. What is the purpose of a splice?

---

3. Name the three intrinsic losses and explain them briefly.
4. Name the three extrinsic losses and explain them briefly.
5. Which would be preferred, a connector or a splice to a computer terminal?
6. What connector would be used for a long-haul system?
7. What is the difference between transmissive and reflective star couplers?

PROBLEMS

1. The core of the receiving fiber can be $\pm 5\%$ of the transmitting fiber, which has a 50 μm core. What is the core mismatch loss?
2. What is the NA loss if the transmitting NA of the core is 0.28 and the NA of the receiving fiber is 0.32?
3. Find the Fresnel loss in decibels for two glass fibers separated by air. The index of refraction of glass is equal to 1.46. The index of refraction for air is 1.0.
4. If the insertion loss of the port is 0.75 dB, what is the ratio of the power output of the port $P_{i,j}$ to the power input to the ports?
5. What would be the splitting loss among eight ports?

6

Optical Sources

CHAPTER OBJECTIVES

The student will be able to:

- Evaluate the use of a particular source in a system.
- Describe the semiconductor physics of the light emitting diode and insertion laser diode.
- Know the advantages and disadvantages of the light emitting diode and injection laser diode.

6.1 INTRODUCTION

A radiation source is used to convert the electrical signal to light to send via the fiber optic cable. The electrical signal can either be *analog*, when the radiation source is intensity modulated, or *digital*, when the source is turned off or on. All of the sources must meet certain criteria:

1. Must have a small emissive surface and be able to couple to the fiber core.
2. Must emit light within the acceptance cone.
3. Must emit at the correct wavelength, the one most advantageous for the fiber to propagate the signal.
4. Must have a rapid response time.
5. Must be reliable and economical.
6. Must be intense.

Two commonly used light emitters for fiber optics are the light emitting diode (LED) and the injection laser diode (ILD). These emitters provide the small size, low voltage, and the desired wavelengths needed for a fiber optic link. This chapter explains the light emitting process briefly and provides performance information on light emitters.

6.2 CREATION OF A PHOTON

Three methods are available to create a photon in a semiconductor: absorption, spontaneous emission, and stimulated emission. Each method is shown in Figure 6-1. In the figure, a band diagram is used to show the valence band (VB) and the conduction band (CB). The distance between the valence and the conduction bands is known as the bandgap or energy gap (E_g). This is the amount of energy needed for the electron to jump from the valence band (equilibrium or ground state) to the conduction band (excited state).

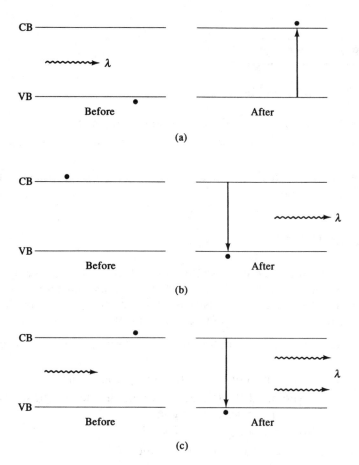

Figure 6-1 (a) Absorption; (b) spontaneous emission; (c) stimulated emission.

Absorption occurs when a photon in the bandgap region gives up its energy to an electron in the valence band. The electron is now in the conduction band because of the increase in energy. Back in the valence band, the electron has left a hole.

Spontaneous emission takes place when an excited electron wants to return to the valence band. The charged electron is in the conduction band and falls back to the valance band. The electron must lose energy to go back to the valence band. The energy that is lost is in the form of a photon. The electron recombines with a hole. This is the process by which the LED produces the photons.

Stimulated emission occurs when the electron is in the excited state (conduction band) with a lot of excess energy. Another external photon emerges within the bandgap. The electron wants to return to the valence band and release a photon as it returns. The external photon stimulates the electron to drop to the valence band, releasing a photon with the same energy as the external photon. Since the photon has the same energy, the same wavelength of light is emitted as the external photon. These photons go on to stimulate more electrons, which create more photons—hence the name "stimulated emission." Stimulated emission is produced only if there are more electrons in the conduction band (excited state) than in the ground state. This is called *population inversion*. A laser (an acronym for "light amplification by stimulated emission of radiation" can produce light only when population inversion is created.

The light must be confined to a narrow region. If the light is not confined, it will emit in every direction. A resonant cavity can be thought of as the container where stimulated emission occurs. It confines the light, making it pass back and forth within this cavity. By bouncing the light from end to end, more stimulated emissions are occurring at a particular wavelength. Finally, one wavelength of light builds up and is released. The light is monochromatic or has only one wavelength. This cavity can be as small as a few centimeters or as long as a few hundred meters.

6.3 THE LIGHT EMITTING DIODE

The light emitting diode (LED) is an excellent device to use in a fiber optic link because of its long life span (10^6 hours), operational stability, wide temperature range, and low cost (some as little as $5). For bit rates less than 50 Mbps, LEDs are the best choice. These devices have good output power, which can be modulated by varying the input current to the device.

The LED is an *electroluminescent device*, which means that to get the light out, current must flow through the device. Electroluminescence occurs when light is generated by the recombination of electrons and holes. Recombination occurs inside a forward-biased *pn* junction of a diode, thereby generating photons. Starting with a *pn* junction, electrons have a high population in the *N*-type material. The holes have a high population in the *P*-type material. Under the influence of an electric field, some of the electrons are injected across the *pn* junction and recombine with the holes. When this recombination occurs, energy is lost and that energy is released in the form of a photon. This is known as *spontaneous emission*. The photons radiate out of the device at different wavelengths, which is known as *incoherent radiation*.

The wavelength of the light emitted from the LED depends on the material's energy gap (bandgap) between the valence band and the conduction band. This bandgap energy of the *pn* junction is defined by the following equation:

$$\lambda = \frac{h \cdot c}{E_g} \tag{6-1}$$

λ is the wavelength of operation

h is Planck's constant, 6.62×10^{-34} joule-second

c is the speed of light

E_g is the bandgap energy, in joules

Because the voltage is relatively small, the bandgap energy is described with the units of electron volts. The following formula converts the bandgap energy to electron volts:

$$\text{bandgap (eV)} = \frac{E_g}{e}$$

$e = 1.602 \times 10^{-19}$ coulomb

The bandgap energies in electron volts (eV) for the following materials are:

Germanium	0.66
Silicon	1.16
Gallium arsenide	1.43
Indium arsenide	0.33
Indium phosphide	1.29
Indium gallium arsenide	0.36–1.43
Indium gallium arsenide phosphide	0.36–1.35

Example 6-1

At 880 nm, what is the bandgap, and what material should be used?

SOLUTION

$$E_g = \frac{6.62 \times 10^{-34} \cdot 3 \times 10^8}{880 \times 10^{-9}} = 2.256 \times 10^{-19}$$

$$= \frac{2.256 \times 10^{-19}}{1.6 \times 10^{-19}} = 1.4 \text{ eV}$$

> The bandgap energy suggests that gallium arsenide be used as the material at 880 nm.

6.3.1 Source Materials

The basic *pn* junction semiconductor material and the added dopants determine the bandgap energy of the emitter and how many photons are given off. The most important materials used are the group III–V materials. They are so called because they come from the group III elements (such as aluminum, gallium, or indium) and group V materials (such as phosphorus or arsenic) of the periodic chart. Various ternary (three-element) and quaternary (four-element) combinations of these materials become good candidates for the device. For instance, a gallium arsenide (GaAs) diode emits peak power at 940 nm with a bandgap of 1.43 eV. Add a small percentage of the dopant aluminum to the device, and a ternary diode of gallium–aluminum–arsenide (GaAlAs) has been constructed. The peak power out is at 800 to 900 nm (infrared radiation), which makes this device emit a usable wavelength for fiber optic transmission. Adding another dopant, phosphorus, making this a quarternary device, can give a peak power output from 1 to 1.7 μm.

6.3.2 Structure of the LED

The device structures for the LED and ILD are very different. The typical device structure for an LED is depicted in Figure 6-2. This is the simplest of the LED structures. It uses a homojunction structure, which means that a single material is used to form the *p* and *n* regions of the chip, such as GaAs. This particular diode has an *n*-type GaAs substrate with a *p*-type GaAs layer on the top side. The light is generated in the *p*-area and is limited by the diode's physical construction. The emitting area is large because the photons are not confined to a narrow region but are radiating from the planar surface. These devices, called surface-emitting diodes, are considered Lambertian

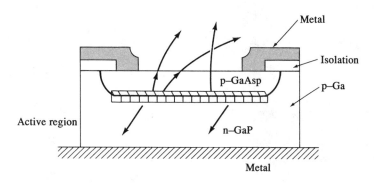

Figure 6-2 LED device structure.

sources, radiating in all directions. The output of the light is then 180°. This wide output pattern makes it difficult to couple more than a small part of the total output into the fiber.

Various structures have been made to improve the coupling efficiency to the fiber, such as the Burrus diode and the edge-emitting diode. These are *heterojunctions*, structures that form a junction of two materials with different bandgap energies. The changes in bandgap form a channel or barrier for both the electrons and the holes. The refractive indices of these materials are also different, causing an optic waveguide to form.

The *Burrus diode* reduces material absorption by etching a dip or well into the diode substrate. The well reduces the size of the light-generating area. The fiber is then epoxied into the well in order to accept the light. The circular active area is approximately 50 μm in diameter and up to 2.5 μm deep. A 200 MHz bandwidth can be achieved with this structure. The device structure is shown in Figure 6-3.

The *edge-emitting diode* uses a narrow contact stripe or narrow active region to confine the output light. This is called a heterojunction device because it uses at least three layers. The upper and lower layers have a lower index of refraction that sandwiches the center layer. This center layer is a stripe that can confine the light into a narrow region. This allows more power, 125 μW, to be coupled into the fiber. The edge-emitting diode can handle bandwidths greater than 200 MHz. The device is shown in Figure 6-4.

6.3.3 Quantum Efficiency

The quality of the conversion of the electrical current to light is known as the *quantum efficiency*. This number describes the number of electrons emitted per unit time, compared to the number of charge carriers crossing the *pn* junction in the semiconductor diode. The excess carriers in the junction can combine radiatively, causing a photon with energy hc/λ to be emitted. If the carriers are combined nonradiatively, energy is released in the form of heat because of crystal lattice vibrations. The excess carriers have a lifetime that

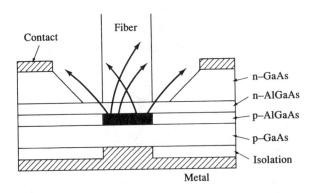

Figure 6-3 Burrus diode.

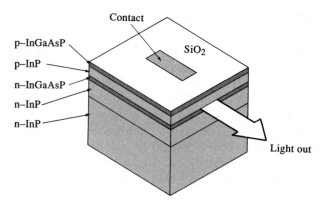

Contact

p–InGaAsP

p–InP

n–InGaAsP

n–InP

n–InP

SiO$_2$

Light out

Figure 6-4   Edge-emitting diode.

decays exponentially with time. The equation is

$$\Delta n = \Delta n_0 e^{-t/\tau} \tag{6-2}$$

Δn_0 the initial injected density

τ is the carrier lifetime

The internal quantum efficiency in the active region is the fraction of excess carriers or electron–hole pairs that combine radiatively during their lifetime. It is defined by the equation

$$\eta_i = \frac{\tau_{nr}}{\tau_{nr} + \tau_r} \tag{6-3}$$

where τ, the bulk recombination lifetime, is

$$\frac{1}{\tau} = \frac{1}{\tau_r} + \frac{1}{\tau_{nr}} \tag{6-4}$$

τ_r the radiative lifetime

τ_{nr} is the nonradiative lifetime

Example 6-2

A silica LED has a nonradiative lifetime of 1×10^{-7} second and a radiative lifetime of 1×10^{-3} second. Calculate the internal quantum efficiency.

SOLUTION

$$\eta_i = \frac{1 \times 10^{-7}}{1 \times 10^{-7} + 1 \times 10^{-3}} = 1 \times 10^{-4}$$

This means that one recombination in every 10,000 gives off a photon.

6.4 THE INJECTION LASER DIODE

The injection laser diode (ILD) is used in fiber optic links primarily because they are capable of producing 10 dB more power than an LED. The ILD produces coherent or monochromatic light. The ILD, below a certain current threshold, has the same characteristics as the LED. At this threshold, spontaneous emissions occur and light is radiated in a wide area. Above this threshold current, monochromatic light is ejected out of the device, and lasing will begin. It is the lasing action that produces the gain. The result is that the ILD will have high optical gain and narrow spectrum or wavelength and coherence. *Coherence* means that all the light emitted is of the same wavelength.

The light-generating process is different in an ILD. Figure 6-5 is a schematic of a basic laser diode. When the ILD is biased at a certain current, the holes and electrons move into the active region. The active region is enclosed by a material that has a higher refractive index than itself, which provides optical guidance and confinement of the electron carriers. Because the active region is sandwiched between layers, emitted light is confined and can only exit from the back and front faces of the ILD device.

Some of the holes and electrons recombine, giving off photons or light. Others are trapped between the end walls of the active region. The walls act like mirrors or a resonant cavity in which the photons are reflected, making many trips from end to end. The photon stimulates the electrons to recombine with the hole, causing a photon to be released at the same wavelength as before. Therefore, the first photon stimulated the emission of the second photon, causing gain to occur.

The ILD can use the same materials as LEDs but the device structure is

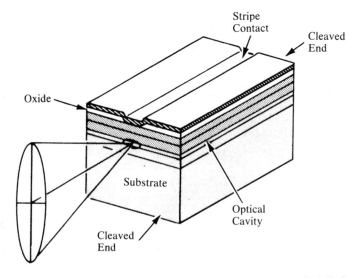

Figure 6-5 Schematic of basic injection laser diode. (Courtesy of AMP, Inc.)

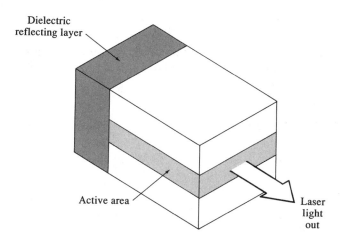

Dielectric
reflecting layer

Active area

Laser
light
out

Figure 6-6 Gain-guided ILD.

different. ILD devices sandwich the active region between layers of higher energy bandgaps and lower refractive index to provide a waveguide for the optical cavity. There are two structural families that characterize the type of light confinement in the laser. They are the gain-guided and the index-guided laser diodes.

Figure 6-6 shows an oxide-stripe gain-guided ILD. This structure can handle a bandwidth of 500 MHz or greater, primarily for the range 800 to 900 nm. This device structure has a very narrow stripe width (confinement for the waveguide).

Figure 6-7 shows an index-guided laser diode. The structure is devised to sandwich the active layer between two layers of material that have a refractive index different from that of the active region. Other types of ILD structures can have bandwidths of up to 3 GHz.

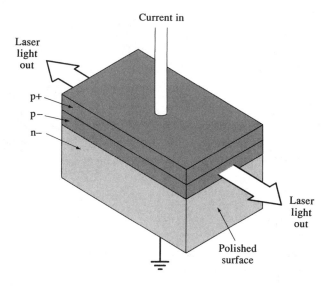

Current in

Laser
light
out

p+
p−
n−

Polished
surface

Laser
light
out

Figure 6-7 Index-guided ILD.

No matter what type of laser diode structure is used, practical operation of the diode requires that it be housed in a hermetically sealed enclosure. The diode is cooled, mitered, and has some type of pigtail that matches the transmission fiber. A pigtail is a short piece of fiber that is coupled to the emitting surface and fed out of the sealed module.

6.5 CHARACTERISTICS OF THE LED AND THE ILD

Similar characteristics exist in the LED and ILD, although their numerical values may be different. The fiber transmission windows, power output, spectral width, and power coupled to the fiber are discussed below.

6.5.1 Transmission Windows

One main characteristic of an optical emitter is that it must be compatible with the fiber transmission windows. The fiber transmission windows occur at 0.85, 1.3, and 1.5 μm for a typical fiber, as shown in Figure 6-8. These windows occur because of the relative lack of hydrogen ion (OH^-) absorption in the fiber, which in turn means that there will be less loss in the fiber. The optical source material can be made to exploit these windows.

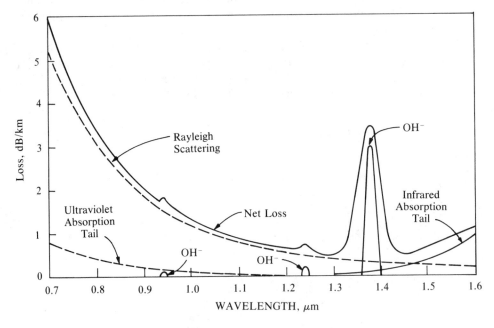

Figure 6-8   Fiber optic transmission window.*

* From Allen H. Cherin, *An Introduction to Optical Fibers for Engineering and Physics*, copyright © 1983 by McGraw-Hill, Inc. Reproduced with permission.

The ratio of the dopants to the material determines the bandgap of the alloy and then in turn the wavelength of the peak power radiation. The following equation is used to find the peak emission wavelength knowing the bandgap energy:

$$\lambda(\mu m) = \frac{1.240}{E_g(eV)} \qquad (6\text{-}5)$$

λ is the wavelength of operation

E_g is the bandgap energy, in electron volts

The ternary $Ga_{1-x}Al_xAs$ is used for operation in the range 800 to 900 nm. By varying the ratio x of the aluminum arsenide to the gallium arsenide, an LED can emit at a particular wavelength in the transmission window of the specified fiber. For example, in the alloy $Ga_{0.93}Al_{0.07}As$, the bandgap energy is 1.51 eV and emits at λ equal to 0.82 μm. The quarternary alloy $In_{0.74}Ga_{0.26}As_{0.56}P_{0.44}$ has a bandgap energy of 0.96 eV, which emits at 1.3 μm.

6.5.2 Power Output

Another concern is the output characteristics of the LED and ILD. The graph in Figure 6-9 shows power output versus applied current. Notice that the LED has a nearly linear dependence of the optical power to the applied current. This means that the LED can be operated at a lower power level consumption. The relationship between optic power and current is

$$P = N \cdot \frac{hc}{\lambda} = \frac{\eta \cdot E_g \cdot I}{e} \qquad (6\text{-}6)$$

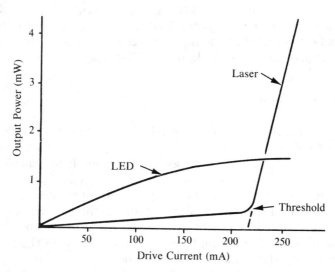

Figure 6-9 Power output versus current for LED and ILD. (Courtesy of AMP, Inc.)

N is the number of charges per second $= i/e$

η is the fraction of radiative charges that recombined

E_g is the bandgap energy, in joules

This formula is based on the internal radiant flux and the quantum efficiency of the radiation device. LEDs benefit from the lower power consumption by having higher reliability and longer life. They operate around 50 to 300 mA and require a voltage of 1.5 to 2.5 V.

Example 6-3

A light emitting diode is made of GaAs, the quantum efficiency is 0.87 and the energy gap is 1.43 eV. The diode has a forward current of 100 mA. Calculate the radiant flux density emitted from the LED. What is the applied voltage of the LED?

SOLUTION

$$P = \frac{0.87 \cdot (1.43 \cdot 1.6 \times 10^{-19}) \cdot 100 \times 10^{-3}}{1.6 \times 10^{-19}} = 124 \text{ mW}$$

The applied voltage is equal to the bandgap energy. Therefore, the applied voltage to the LED is $V = E_g(\text{eV}) = 1.4$ V.

The power output of an ILD is small at low currents since only spontaneous emissions are occurring. As the applied current is increased, the spontaneous emissions become stimulated emissions. The output power increases dramatically. The level of the input power in the lasing mode must be very carefully controlled or the diode will be destroyed by heat.

6.5.3 Spectral Width

The light spectra of the LED and the ILD are quite different. Neither device outputs a fixed wavelength of light. The output wavelength is based on a number of factors (especially energy bandgap) and is measured as a function of output power. Spectral width is measured at the half-power points (or the full-width half-maximum point) on a graph of output power versus emitted wavelength. The difference of wavelengths at the two half-power points is called the *spectral width*. Light emitters are specified to operate at a certain center wavelength. For example, if an emitter has a 1300 nm center wavelength, in reality the emitter does not produce a constant wavelength. Instead, one or more wavelengths that are within 10 to 20 nm of the center point are produced.

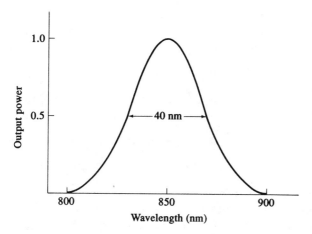

Figure 6-10 Spectrum of the LED.

An LED with a specified wavelength of 850 nm will have a spectral width of 30 to 50 nm. High spectral widths are associated with the chromatic dispersion type of pulse broadening, which limits the bandwidth of the fiber optic system. See Figure 6-10 for the wavelength spectrum of an LED.

A 1300 nm ILD typically has a lasing spectral width of 0.01 nm. The materials used within the ILD affect its spectral width. ILD devices made from the ternary compound gallium aluminum arsenide (GaAlAs) and the quaternary indium gallium arsenide phosphide (InGaAsP) have nonlasing spectral widths of 2 to 5 nm and 0.8 nm, respectively. The nonlasing spectral widths occur when the ILD is acting like an LED. See Figure 6-11 for the spectrum of an ILD. Note the noise present at other wavelengths for the ILD. LED systems do not have this noise associated with them.

6.5.4 Power Coupled to the Fiber

An LED can typically couple 10% of its power (10 to 100 μW) into a 100-μm-core-diameter step-index fiber (Figure 6-12). Approximately 1% of the power

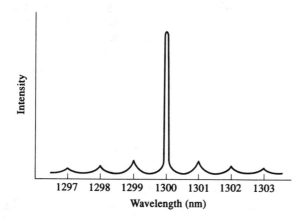

Figure 6-11 Spectrum of the ILD.

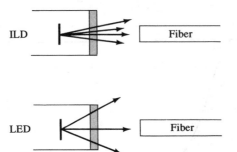

Figure 6-12 Power coupled into fiber by LED and ILD.

can be coupled into a 50 μm graded-index fiber. These low coupling efficiencies are due to the wide angle beam spread characteristic of an LED. In contrast, the double heterostructure ILD described earlier has a coupling efficiency of 50% into a 50 μm graded-index fiber. This means that 1 to 10 mW of optical power is coupled into the fiber.

6.5.5 Speed

The source must be able to turn off and on fast enough for the bandwidth of the system (Figure 6-13). The speed is usually specified by the risetime of the source. LED risetime is from 3 to 100 ns and the laser diode has a risetime of less than 1 ns. The bandwidth of the source can be approximated from the risetime using the formula

$$B = \frac{0.35}{t_r} \tag{6-7}$$

B is the bandwidth of the source

t_r is the risetime, in seconds

6.5.6 Reliability and Lifetime

LEDs are more reliable than laser diodes because they are less sensitive to device degradation. Over a period of time, both will degrade because of crystal lattice defects, which may cause failures. High current surges cause failures in laser diodes. An LED has a longer expected lifetime than that of a laser diode. The laser diode has a warranted lifetime of 10,000 to 100,000 hours, whereas an LED has 10 times that great a lifetime.

6.6 SUMMARY

Table 6-1 summarizes the similarities and differences between the LED and the ILD. The LED is used for short, lower-bandwidth systems. The ILD is used for almost any other bandwidth system. Included in the table are the bit rate and the transmission distance for the devices. For instance, if using an

TABLE 6-1 LED/ILD Comparison

Characteristic	LED		ILD	
Material	III–V compounds		III–V compounds	
Output	Incoherent		Coherent	
Response time (ns)	2–20		0.01–1.0	
Spectral width rms (nm)	15–60		0.2–5	
Power coupled into the fiber (μW)	10–100		1–10	
Wavelength (nm)	880–1550		880–1550	
Risetime, 10–90% (ns)	2–20		≤ 1	
Voltage drop (V)	1.5–2.5		1.5–2.0	
Forward current (mA)	50–300		10–300	
Threshold current (mA)	NA		5–250	
Transmission distance (km) at bit rate (Mbps)	km	Mbps	km	Mbps
	0.01–0.1	5–10	5–20	≤ 565
	1–5	30–100		
	1–5	50–200	35	≤ 1200

LED at a non-return-to-zero bit rate (digital) of 10 Mbps coupled into a 50-μm fiber, the transmission distance (the distance that the light can be detected) would be approximately 0.1 km.

6.7 EQUATION SUMMARY

Wavelength in terms of bandgap energy:

$$\lambda = \frac{h \cdot c}{E_g} \tag{6-1}$$

Number of excess carriers:

$$\Delta n = \Delta n_0 e^{-t/\tau} \tag{6-2}$$

Internal quantum efficiency:

$$\eta_i = \frac{\tau_{\text{nr}}}{\tau_{\text{nr}} + \tau_{\text{r}}} \tag{6-3}$$

Bulk recombination lifetime:

$$\frac{1}{\tau} = \frac{1}{\tau_{\text{r}}} + \frac{1}{\tau_{\text{nr}}} \tag{6-4}$$

Peak emission wavelength:

$$\lambda(\mu\text{m}) = \frac{1.240}{E_g(\text{eV})} \tag{6-5}$$

Fiber Optics
Infrared LED

MFOE200

HERMETIC FAMILY
FIBER OPTICS
INFRARED LED

... designed as an infrared source in low frequency, short length Fiber Optics Systems.
   Typical applications include: medical electronics, industrial controls, M6800 Microprocessor systems, security systems, etc.

● High Power Output Liquid Phase Epitaxial Structure
● Performance Matched to MFOD100, 200, 300
● Hermetic Metal Package for Stability and Reliability
● Compatible With AMP Mounting Bushing #227015

CASE 209-02
METAL

MAXIMUM RATINGS

Rating	Symbol	Value	Unit
Reverse Voltage	V_R	3	Volts
Forward Current — Continuous	I_F	60	mA
Total Device Dissipation @ T_A = 25°C Derate above 25°C	$P_D(1)$	250 2.27	mW mW/°C
Operating Temperature Range	T_A	−55 to +125	°C
Storage Temperature Range	T_{stg}	−65 to +150	°C

ELECTRICAL CHARACTERISTICS (T_A = 25°C)

Characteristics	Symbol	Min	Typ	Max	Unit
Reverse Leakage Current (V_R = 3 V, R_L = 1 Megohm)	I_R	—	50	—	nA
Reverse Breakdown Voltage (I_R = 100 μA)	$V_{(BR)R}$	3	—	—	Volts
Forward Voltage (I_F = 100 mA)	V_F	—	1.5	1.7	Volts
Total Capacitance (V_R = 0 V, f = 1 MHz)	C_T	—	18	—	pF

OPTICAL CHARACTERISTICS (T_A = 25°C)

Characteristics		Symbol	Min	Typ	Max	Unit
Total Power Output (2) (I_F = 100 mA, λ ≈ 940 nm)	See Figures 1 and 2	P_O	2	3	—	mW
Power Launched (3) (I_F = 100 mA)		P_L	35	45	—	μW
Optical Turn-On and Turn-Off Time		—		t_{on}, t_{off}	250	—	ns

(1) Printed Circuit Board Mounting
(2) Total Power Output, P_O, is defined as the total power radiated by the device into a solid angle of 2π steradians.
(3) Power Launched, P_L, is the optical power exiting one meter of 0.045" diameter optical fiber bundle having NA = 0.67, Attenuation = 0.6 dB/m @ 940 nm,
   terminated with AMP connectors. (See Figure 1.)

Figure 6-13   Specification sheets for sources. (Courtesy of Motorola, Inc.)

MFOE200

TYPICAL CHARACTERISTICS

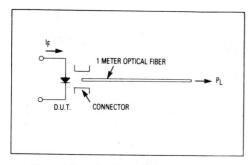

Figure 1. Launched Power Test Configuration

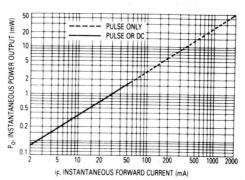

Figure 2. Instantaneous Power Output
versus Forward Current

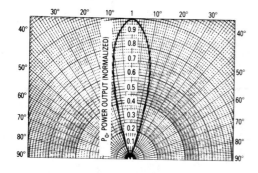

Figure 3. Spatial Radiation Pattern

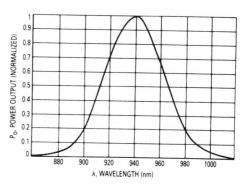

Figure 4. Relative Spectral Output

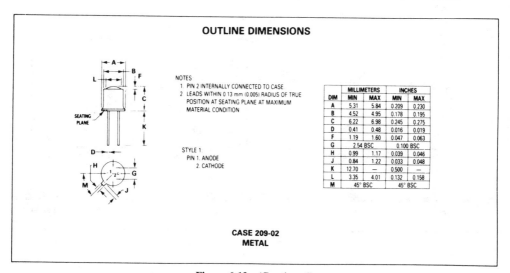

OUTLINE DIMENSIONS

NOTES
1. PIN 2 INTERNALLY CONNECTED TO CASE
2. LEADS WITHIN 0.13 mm (0.005) RADIUS OF TRUE
   POSITION AT SEATING PLANE AT MAXIMUM
   MATERIAL CONDITION

STYLE 1
PIN 1. ANODE
    2. CATHODE

DIM	MILLIMETERS		INCHES	
	MIN	MAX	MIN	MAX
A	5.31	5.84	0.209	0.230
B	4.52	4.95	0.178	0.195
C	6.22	6.98	0.245	0.275
D	0.41	0.48	0.016	0.019
F	1.19	1.60	0.047	0.063
G	2.54 BSC		0.100 BSC	
H	0.99	1.17	0.039	0.046
J	0.84	1.22	0.033	0.048
K	12.70	—	0.500	—
L	3.35	4.01	0.132	0.158
M	45° BSC		45° BSC	

CASE 209-02
METAL

Figure 6-13   (Continued)

Sec. 6.7   *Equation Summary*

Order this data sheet
by MFOE76/D

MOTOROLA
■ **SEMICONDUCTOR** ■
TECHNICAL DATA

Fiber Optics — FLCS Family
Visible Red LED

This device is designed for low cost, medium frequency, fiber optic systems using 1000 micron core plastic fiber. It is compatible with Motorola's wide variety of detector functions from the MFOD70 series. The MFOE76 employs gallium aluminum technology, and comes pre-assembled into the convenient and popular FLCS connector.

- Low Cost
- Very Simple Fiber Termination and Connection. See Figure 9
- Convenient Printed Circuit Mounting
- Integral Molded Lens for Efficient Coupling
- Mates with 1000 Micron Core Plastic Fiber, such as Eska SH4001

MFOE76

FLCS FAMILY
FIBER OPTICS
VISIBLE RED
LED
660 nm

CASE 363B-01

MAXIMUM RATINGS

Rating	Symbol	Value	Unit
Reverse Voltage	V_R	5	Volts
Forward Current — Continuous	I_F	60	mA
Forward Current — Peak Pulse	I_F	1	A
Total Power Dissipation @ T_A = 25°C (1) Derate above 35°C	P_D	132 2	mW mW/°C
Ambient Operating Temperature Range	T_A	− 40 to + 100	°C
Storage Temperature	T_{stg}	− 40 to + 100	°C
Lead Soldering Temperature (2)	—	260	°C

Notes: 1. Measured with device soldered into a typical printed circuit board.
2. 5 seconds max; 1/16 inch from case.

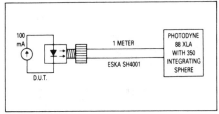

Figure 1. Power Launched Test Setup

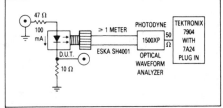

Figure 2. Optical Turn-On and Turn-Off Test Setup

Ⓜ **MOTOROLA** ■

DS2740

Figure 6-13 (Continued)

Optical power:

$$P = N \cdot \frac{hc}{\lambda} = \frac{\eta \cdot E_g \cdot I}{e} \qquad \text{(6-6)}$$

Bandwidth dependent on how the signal is sent (i.e., NRZ, Manchester):

$$B = \frac{0.35}{t_r} \qquad \text{(6-7)}$$

QUESTIONS

1. Name six criteria for an optical source.
2. Name three ways to create or use a photon and explain them.
3. What are some advantages in using a light emitting diode?
4. Why is quantum efficiency important to optical source?
5. Why are injection laser diodes used for fiber optics?
6. What are the three fiber optic cable transmission windows?

PROBLEMS

1. A material has a bandgap energy of 1.78 eV. What would be the wavelength for the material?
2. If the peak wavelength for an InGaAsP semiconductor is 1.35 μm, what is its bandgap energy?
3. What should be the approximate bandgap energies at the fiber optic transmission windows (0.9, 1.3, and 1.5 μm) be?
4. What is the bandwidth of an LED with a risetime of 5 ns?
5. In an n-type GaAsP semiconductor, the radiative lifetime of the electron is 10^{-9} second and its nonradiative recombination is 10^{-8} second. Find the internal quantum efficiency and the bulk recombination lifetime for the semiconductor.
6. An ILD is being used to couple light into a fiber optic cable. The ILD has a 50% coupling efficiency into a 50 μm graded-index fiber and that 5.5 mW of power has been coupled in. What should the initial power out of the fiber be?
7. If the risetime of a LED is 4.35 ns, what is the bandwidth of the source?
8. A bandwidth of 40 MHz is needed for a fiber optic system. What type of diode should be used, and what would the risetime be?

7

Photodetectors

CHAPTER OBJECTIVES

The student will be able to:

- Evaluate the use of a particular detector in a system.
- Describe the semiconductor physics of the PIN and avalanche photodiodes.
- Understand the different types of noise associated with detectors.
- Know the advantages and disadvantages of the PIN and avalanche photodiodes.

7.1 INTRODUCTION

Photodetectors are used as light receivers for fiber optic communication systems. They convert the optical energy into electrical energy. Since the optical signal is very weak, the sensitivity of the detector is critical for the overall fiber optic link performance. The ideal photodiode must have the following characteristics:

1. Must be sensitive to weak light falling on the device.
2. Must be able to operate in the near-infrared (850, 1300, 1550 nm).
3. Must be sufficiently fast to be able to transform the light to electrons.
4. Must have small dimensions compatible with the fiber.
5. Must be low in cost.
6. Must be insensitive to the environment.

Two commonly used receiver types are the PIN (positive–intrinsic–negative) photodiode and the APD (avalanche photodiode). The light detection process will be explained briefly. The performance information for these two types of photodetectors is discussed in this chapter.

7.2 THE PIN PHOTODIODE

The PIN photodiode is a type of depletion-layer photodiode and is the most commonly used fiber optic detector. The lightly n-doped intrinsic layer is sandwiched between a thin layer of positively doped and negatively doped materials. Figure 7-1 shows a typical structure. Light falls on the p layer of the photodiode, where it is absorbed by an electron. The electron moves toward the depletion region (the intrinsic layer) due to the applied reverse bias voltage. Below a certain wavelength known as the *cutoff wavelength*, the detector will not convert the light to an electrical current. The cutoff wavelength is dependent on the bandgap of the device being used.

$$\lambda_c = \frac{h \cdot c}{E_g} = \frac{1.24}{E_g} \qquad (7\text{-}1)$$

λ_c is the wavelength of operation, micrometers

h is Planck's constant

c is the speed of light

E_g is the bandgap energy of the material, in electron-volt

Example 7-1

Silicon has a bandgap of 1.1 eV. What is the wavelength of the photons that will be absorbed?

SOLUTION

$$\lambda = \frac{hc}{E_g} = \frac{(6.6 \times 10^{-34})(3.0 \times 10^8)}{(1.6 \times 10^{-19})(1.1)} = 1.13 \ \mu\text{m}$$

The detector will respond only to wavelengths higher than 1.13 μm.

The electron must be collected in the conduction band to obtain an electric current or photocurrent. The electron–hole pairs are separated by the aid of the electric field in the intrinsic region. The thickness of the intrinsic region should be at least as wide as the light absorption region (negatively charged region) (Figure 7-2). The thickness is important because it facilitates the probability of the incident photons to create electron–hole pairs. The n region should be very narrow to enable electrons to move out rapidly through the

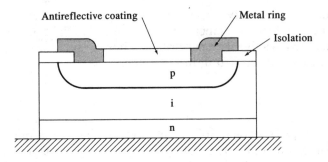

Figure 7-1   PIN structure.

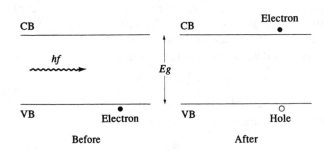

Figure 7-2   Optical absorption coefficient. (Courtesy of Les Editions Le Griffon D'Argile.)

device. The structure of the photodiode determines the speed of response to the incident light and the efficiency at which the light is converted. PIN photodetectors are used for short transmission distances and lower modulation frequencies. They are used as receivers for LEDs. The reverse-bias voltage operates at 7 to 10 volts.

7.3 THE AVALANCHE PHOTODIODE

Avalanche photodiode (APD) devices are operated at high reverse-bias voltages, typically higher than 300 volts, where internal amplification or avalanche multiplication of the photocurrent occurs. APDs are suitable for long transmission distances and high modulation frequencies (100 GHz). They are commonly used with laser and ILD light sources. The commonly used structure is called the *reach-through construction APD* (RAPD). The reach-through avalanche photodiode is composed of a *p*-type material deposited on a p^+ substrate. A *p*-type diffusion is made in the high-resistivity material and a n^+ layer is added as shown in Figure 7-3. The structure is referred to as a $p^+\pi pn^+$. The π layer is an intrinsic layer that has been doped with *p*-type material such as boron or phosphorus. The APD acts like a *pn* junction, at low reverse bias. As the reverse bias voltage is increased, the electric field also increases, causing the depletion region to reach through to the intrinsic layer.

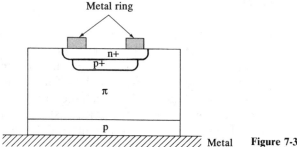

Metal   **Figure 7-3**   APD structure.

Light will enter to the device through the p^+ region and be absorbed in the π material. The photon gives up its energy and creates electron–hole pairs. The electrons drift to the pn^+ junction, where a high electric field exists. This high electric field accelerates the electron–hole pairs, causing collisions with other pairs. This produces additional pairs or multiplication of the carriers. This collision process causes the avalanche multiplication and is defined mathematically below.

$$M = \left(\frac{1}{1 - V/V_B}\right)^n \tag{7-2}$$

V is the reverse voltage applied

V_B is the breakdown voltage

n is a constant between 3 and 6 that depends on material and wavelength

Example 7-2

A typical avalanche photodiode has a breakdown voltage of 300 volts and a reverse voltage of 100 volts. What is the multiplication if the material constant is 5?

SOLUTION

$$M = \left[\frac{1}{1 - (100/300)}\right]^5 = 7.59$$

The multiplication factor is only an average. The answer of 7.59 means that one photon gives up its energy and 7.59 electron–hole pairs are created. This exchange of energy causes multiplication noise.

The average number of electron–hole pairs created by a photon per unit distance is called the *ionization rate*. The ionization rates differ for the electrons and the holes.

7.4 PHOTODIODE PARAMETERS

For the photodiode to interpret the information contained in the optical signal, certain requirements must be met. The photodiode must be sensitive to the emission wavelength range of the optical source being used. It must add little or no noise to the system. For the photodiode to be effective, it must have a fast response time or sufficient bandwidth to handle the data rate. These parameters are needed in the analysis for the total system risetimes, noise, and frequency response. The quantum efficiency, quantum noise, photodiode responsivity, speed of response, dark current, noise equivalent power, and signal-to-noise ratio are the parameters that most manufacturers define. Most system designers need to have this information to design a fiber optic system. The parameters are defined in the following sections.

7.4.1 Quantum Efficiency

The quantum efficiency, η, is the efficiency of the conversion of incoming photons to electron–hole pairs. It gives the average number of electron–hole pairs generated for each incident photon. It is a statistical parameter and describes only a time average of the conversion process. Quantum efficiency depends on the wavelength, detector material, and angle of incidence and increases with the width of the depletion region in the photodiode. The following expression relates the numerical value of η to responsivity and the wavelength of light incident to the material:

$$\eta = \frac{\text{number of electron–hole pairs generated}}{\text{number of incident photons}}$$

$$= \frac{I_p/e}{P_0/h\nu} \tag{7-3}$$

I_p is the average photocurrent

e is the charge of an electron, 1.6×10^{-19}

P_0 is the average optical power incident on the photodetector

$h\nu = hc/\lambda =$ Planck's constant times the speed of light divided by the wavelength

In an average photodiode, 100 photons will create between 30 and 95 electron–hole pairs. Thus the quantum efficiency values will range from 30 to 95%, depending on wavelength and detector type. A higher quantum efficiency requires that the depletion layer be thick enough to permit a large fraction of the light to be absorbed. The thickness must not be large because it will take longer for the carriers to drift across the reverse-bias junction, thereby causing the speed of response to be slower.

7.4.2 Photodiode Responsivity

Another performance parameter, the responsivity, ρ, is related to quantum efficiency by the following equation:

$$\rho = \frac{I_p}{P_o} = \eta \cdot \frac{e}{h \cdot c} \cdot \lambda \qquad \text{amperes/watt} \qquad (7\text{-}4)$$

I_p is output photocurrent

P_o is input optical power

η is quantum efficiency

e is the charge of an electron, 1.6×10^{-19}

h is Planck's constant

λ is wavelength

This equation compares the output photocurrent in amperes to the input optical power in watts. There are several forms of this equation, all of which lead to typical values for responsivity of 0.65 μA/μW for silicon at 800 nm, 0.45 μa/μW for germanium at 1.3 μm, and 0.6 μA/μW for InGaAs at 1.3 μm.

Example 7-3

A photodiode has a responsivity of 0.62 A/W at a wavelength of 0.9 μm. Its reverse-bias voltage is 20 V. The surface of the photodiode receives a radiant flux of 5×10^{-6} W. What is the quantum efficiency? What is the output photocurrent?

SOLUTION

$$I_p = S_d \cdot P_o = (0.62)(5 \times 10^{-6}) = 3.1 \ \mu\text{A}$$

$$\eta = S_d \cdot \frac{hc}{\lambda} \cdot \frac{1}{e} = 0.86 = 86\%$$

The result shows that 86% of the light that falls on the detector is absorbed or converted to electrons.

7.4.3 Dark Current

Dark current is the unwanted noise that occurs when current flows and there is no input signal. The amplitude of this signal, although in the nanoampere range, can cause problems in the overall output signal. In a typical optical receiver, a preamplifier is added to amplify the current out of the photodiode. Even when the dark current signal is in the nanoampere region, the preamp does not know if it is the true signal or noise. The dark current is caused by surface leakage or bulk leakage currents within the diode. The magnitude of the dark current (i_d) is

$$i_d = 2 \cdot e \cdot I_d \cdot M^2 \cdot F_d \cdot B \qquad \text{amperes} \qquad (7\text{-}5)$$

e is the charge on the electron

I_d is the noise component of the dark current

M is the avalanche gain of the APD ($M = 1$ for PIN)

F_d is the excess noise factor for an APD ($F_d = 1$ for PIN)

B is the bandwidth

Dark current is dependent on the applied voltage, the operating temperature, and the avalanche gain of the diode. As an example, at 25°C, a silicon APD with a 0.1 mm diameter active area has a dark current of 1 to 10 nA with a multiplication of 100. Germanium APDs have a higher dark current, typically 1 μA for $M = 20$.

7.5 DETECTOR NOISE

Noise will always be a problem in a transmission system. It is the unwanted electrical or optical energy exchange that overshadows the signal. The optical signal is very low (nanowatts) and the noise from the energy conversion can be the same or larger than the signal. Once the signal is converted to electrons, the electronic signal needs to be amplified. At the amplifier stage, both the signal and the noise are amplified.

7.5.1 Quantum Noise

Quantum noise, which is related to quantum efficiency, is a measure of the variation of the instantaneous conversion of photons to electron–hole pairs. The evaluation of quantum noise depends on the statistical number of electrons necessary to detect a single light pulse in an ideal photodetector. Practical systems operate at about 400 photons per pulse.

7.5.2 Shot Noise

Shot noise is caused by the random process of the photon creating electron–hole pairs, which in turn create the current.

$$i_{sn}^2 = 2eiB \qquad (7\text{-}6)$$

e is the charge of an electron, 1.6×10^{-19}

i is average current, including dark and signal

B is the bandwidth of the receiver

From the equation, shot noise increases with bandwidth and current.

Example 7-4

If the dark current is 25 nA and the receiver bandwidth is 10 MHz, what is the shot noise?

SOLUTION

$$i_{sn}^2 = 2(1.6 \times 10^{-19})(25 \times 10^{-9})(10 \times 10^6) = 8 \times 10^{-20}$$

$$i_{sn} = 282.8 \text{ pA}$$

Shot noise can occur even when there is no light falling on the photodetector.

7.5.3 Thermal Noise

The noise associated with current that has been generated thermally is called *thermal noise*. Thermal noise (also called Johnson noise or Nyquist noise) arises from the fluctuations in the load resistor of the photodetector. The thermal energy from the load resistor allows the electrons to move randomly, which creates a random current.

$$I_{th}^2 = \frac{4K \cdot T \cdot B}{R_L} = i_d B \qquad (7\text{-}7)$$

K is Boltzmann's constant, 1.38×10^{-23} J/K

B is the bandwidth of the receiver

T is the temperature, in kelvin

R_L is the value of the load resistor

Example 7-5

If the photodetector is at a temperature of 298°K, the load resistor is 200 MΩ, and the bandwidth is 10 MHz, what is the thermal noise?

$$i_{th}^2 = \frac{4(1.38 \times 10^{-23})(298)(10 \times 10^6)}{2 \times 10^8} = 822.48 \times 10^{-24}$$

$$i_{th} = 28.68 \times 10^{-12} = 28.68 \text{ pA}$$

7.5.4 Total Noise

The total noise, i_n, is the square root of the sum of the squares of the shot noise and thermal noise. The equation is

$$i_n = (i_{sn}^2 + i_{th}^2)^{1/2} \qquad (7\text{-}8)$$

Example 7-6

Find the total noise in Examples 7-4 and 7-5.

SOLUTION

$$i_n = \sqrt{(282.8 \times 10^{-12})^2 + (28.68 \times 10^{-12})^2}$$
$$= 284.25 \times 10^{-12} = 284.25 \text{ pA}$$

As can be seen, the total noise current is approximately equal to the shot noise.

7.6 SPEED OF RESPONSE

The speed of response, t_d, for a photodiode depends on the type of material that is to be used. Photons can be absorbed only over a certain spectral wavelength at a certain speed that is constrained by the type of material used for the photodiode. Besides the spectral and frequency response factors, the bulk resistance and capacitance effects in the diode affect both the response time and the noise. The speed of response is also affected by the RC time constant of the external circuitry attached to the diode.

$$\text{speed of response} = t_d = \frac{x^2}{D} \qquad (7\text{-}9)$$

x is the distance between the absorption and the transition layer

D is the diffusivity of the carrier (semiconductor parameter)

Therefore, the cutoff frequency of the photodiode is affected by the speed of response: that is,

$$f_c = \frac{2.8}{2\pi t_d} \qquad (7\text{-}10)$$

7.7 SIGNAL-TO-NOISE RATIO AND DETECTIVITY

The signal in any system must always be higher than the noise level. For an optical signal, it should be twice the noise current. If the signal strength is at about the same level as the noise, the signal will be lost and indistinguishable from the noise. The *signal-to-noise ratio* (SNR) expresses how much stronger the signal is than the noise. A large SNR means that the signal is much higher than the noise. The following equation defines SNR.

$$\text{SNR} = \left(\frac{\eta \cdot e}{h \cdot f}\right)^2 \cdot \frac{(M \cdot m \cdot P)^2}{i_n} \qquad (7\text{-}11)$$

η is the quantum efficiency

e is the charge of an electron, 1.6×10^{-19}

h is the Planck's constant

f is the frequency of the signal

M is the avalanche multiplication

m is the optical modulation index of the incident signal

P is the incident optical power

i_n is the total noise current, consisting of the quantum noise, dark current noise, bulk leakage current, and the noise of the preamplifier

A SNR required for a telephone voice channel is less than that required for a television signal. This is because the ear cannot distinguish frequency distortion as well as the eye can detect noise distortion in the form of picture quality. Broadcast quality television produced by the networks has a higher SNR than that of the television signal. During transmission and reception, there is a high probability that noise will also be picked up in the system. Therefore, the SNR must be higher to start with.

Detectivity, D, is the signal-to-noise ratio that is normalized for the photodiode active area (the part that is illuminated with the light) and the bandwidth of the photodiode. The following equation defines D^*:

$$D^* = \frac{(A \cdot B \cdot \rho)^{1/2}}{i_n} \qquad (Hz/W)^{1/2} \qquad (7\text{-}12)$$

A is the photodiode active area, in square centimeters

B is the operating bandwidth, in hertz

ρ is the responsivity, in amperes/watt

i_n is the total noise current

D^* (rather than SNR) is defined on most specification sheets. Values for D^* range from 10^{11} to 10^{13} $(Hz/W)^{1/2}$ for silicon photodiodes.

Example 7-7

> The speed of response of a certain PIN photodetector is 6 ns, and the quantum efficiency of 0.6 $\mu A/\mu W$ for InGaAs at 1.3 μm. The frequency is 10 MHz. The active area is 1 cm^2, the responsivity is 30 $\mu A/W$, and the total noise current is 18 nA. The incident optical power is 0.379 μW. Find the cutoff frequency, SNR, and D^*.
>
> SOLUTION
>
> $$f_c = \frac{2.8}{2\pi(6 \times 10^{-9})} = 74.27 \text{ MHz}$$

$$\text{SNR} = \left[\frac{(0.6)(1.6 \times 10^{-19})}{(6.626 \times 10^{-34})(10 \times 10^{6})} \right]^{2} \frac{((1)(1)(0.379 \times 10^{-6}))^{2}}{1.8 \times 10^{-9}} = 1.675 \times 10^{9}$$

$$D^{*} = \frac{[(1)(10 \times 10^{6})(30 \times 10^{-6})]^{1/2}}{18 \times 10^{-9}} = 962 \text{ MHz/W}^{1/2}$$

7.8 BIT ERROR RATE

Bit error rate (BER) replaces SNR for a digital system. BER is the ratio of correctly transmitted bits to incorrectly received bits. Typical specifications require a BER of 10^{-9} to 10^{-12}. A ratio of 10^{-9} means that one wrong bit is received for every billion transmitted. For instance, a network of automatic teller machines communicates via digital data. If one bit is lost in a data stream, it may mean that the deposit that was made may not have been recorded. Voice data have a lower BER than that of digital data.

7.9 NOISE EQUIVALENT POWER

The *noise equivalent power* (NEP) is the amount of incident light power for which the signal-to-noise ratio of the photodiode is just equal to unity. It is therefore a measure of the minimum detectable power.

$$\text{NEP} = \frac{V_n}{M \cdot \rho \cdot R} \quad \text{W/Hz}^{1/2} = \frac{\text{total rms output noise current}}{\text{responsivity} \cdot \text{bandwidth}^{1/2}} = \frac{i_n}{\rho} \qquad (7\text{-}13)$$

V_n is the total noise voltage

M is the multiplication of the photodiode

ρ is the responsivity of the photodetector

R is the resistor of input circuit to the photodiode

i_n is the total noise current

NEP values range from 10^{-15} (W/Hz)$^{1/2}$ for diodes with a small active area (low-noise diodes) to 10^{-12} (W/Hz)$^{1/2}$ for large-area-cell diodes.

Example 7-8

Given the total noise current in Example 7-6 and using a responsivity of 30×10^{-6} A/W, find NEP.

SOLUTION

$$NEP = \frac{284.25 \times 10^{-12}}{30 \times 10^{-6}} = 9.475 \times 10^{-6} = 9.475 \ \mu W$$

NEP is sometimes given in W/Hz$^{1/2}$. This is an indication of the amount of power at that particular bandwidth. If the bandwidth is 10 MHz, the NEP is equal to 9.475 μW/(10 Mz)$^{1/2}$, or 2.99 nW/Hz$^{1/2}$.

7.10 SUMMARY

The materials generally used for photodetectors and their specifications are shown in Table 7-1. The most widely used devices are silicon PIN diodes and APDs at optical wavelengths of 0.8 to 0.9 μm, and InGaAs PINs and APDs at wavelengths of 1.1 to 1.6 μm. Figure 7-4 shows a specification sheet for some photodiodes.

7.11 EQUATION SUMMARY

Cutoff wavelength:

$$\lambda_c = \frac{h \cdot c}{E_g} = \frac{1.24}{E_g} \tag{7-1}$$

Avalanche multiplication:

$$M = \left(\frac{1}{1 - V/V_B}\right)^n \tag{7-2}$$

Quantum efficiency:

$$\eta = \frac{\text{number of electron–hole pairs generated}}{\text{number of incident photons}} = \frac{I_p/e}{P_o/h\nu} \tag{7-3}$$

TABLE 7-1 Characteristics of Junction Photodiodes

Material	Structure	Rise-time (ns)	Wavelength (nm)	Responsivity (A/W)	Dark Current (nA)	Gain
Silicon	PIN	0.5	300–1100	0.5	1	1
Germanium	PIN	0.1	500–1800	0.7	200	1
InGaAsP	PIN	<1.0	1000–1700	0.6	15–50	1
Silicon	APD	0.5	400–1000	77	15	150
Germanium	APD	1.0	1000–1600	30	700	50
InGaAsP	APD	0.2	1000–1700	—	100	10–50

Fiber Optics
Photo Detector
Transistor Output

MFOD200

HERMETIC FAMILY
FIBER OPTICS
PHOTO DETECTOR
TRANSISTOR OUTPUT

. . . designed for infrared radiation detection in medium length, medium frequency Fiber Optics Systems.

Typical applications include: medical electronics, industrial controls, security systems, M6800 Microprocessor Systems, etc.

- Spectral Response Matched to MFOE200
- Hermetic Metal Package for Stability and Reliability
- High Sensitivity for Medium Length Fiber Optics Control Systems
- Compatible with AMP Mounting Bushing #227015

CASE 82-05
METAL

MAXIMUM RATINGS (T_A = 25°C unless otherwise noted)

Rating	Symbol	Value	Unit
Collector-Emitter Voltage	V_{CEO}	40	Volts
Emitter-Base Voltage	V_{EBO}	10	Volts
Collector-Base Voltage	V_{CBO}	70	Volts
Light Current	I_L	250	mA
Total Device Dissipation @ T_A = 25°C Derate above 25°C	P_D	250 2.27	mW mW/°C
Operating Temperature Range	T_A	− 55 to + 125	°C
Storage Temperature Range	T_{stg}	− 65 to + 150	°C

STATIC ELECTRICAL CHARACTERISTICS (T_A = 25°C)

Characteristic	Symbol	Min	Typ	Max	Unit
Collector Dark Current (V_{CE} = 20 V, H ≈ 0) T_A = 25°C	I_{CEO}	—	—	25	nA
T_A = 100°C		—	4	—	μA
Collector-Base Breakdown Voltage (I_C = 100 μA)	$V_{(BR)CBO}$	50	—	—	Volts
Collector-Emitter Breakdown Voltage (I_C = 100 μA)	$V_{(BR)CEO}$	30	—	—	Volts
Emitter-Collector Breakdown Voltage (I_E = 100 μA)	$V_{(BR)ECO}$	7	—	—	Volts

OPTICAL CHARACTERISTICS (T_A = 25°C)

Responsivity (Figure 2)	R	14.5	18	—	μA/μW
Photo Current Rise Time, Note 1 (R_L = 100 ohms)	t_r	—	2.5	—	μs
Photo Current Fall Time, Note 1 (R_L = 100 ohms)	t_f	—	4	—	μs

Note 1. For unsaturated response time measurements, radiation is provided by pulsed GaAs (gallium-arsenide) light-emitting diode (λ = 940 nm) with a pulse width equal to or greater than 10 microseconds, I_C = 1 mA peak.

Figure 7-4 Specification sheets for a detector. (Courtesy of Motorola, Inc.)

MFOD200

TYPICAL CHARACTERISTICS

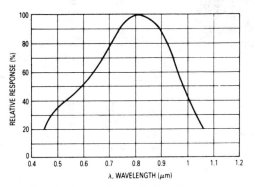

Figure 1. Constant Energy Spectral Response

Figure 2. Coupled System Performance
versus Fiber Length*

*0.045" Dia. Fiber Bundle, N.A. ≅ 0.67,
Attenuation at 940 nm ≅ 0.6 dB/m

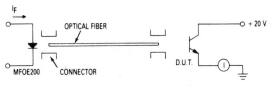

Figure 3. Responsivity Test Configuration

OUTLINE DIMENSIONS

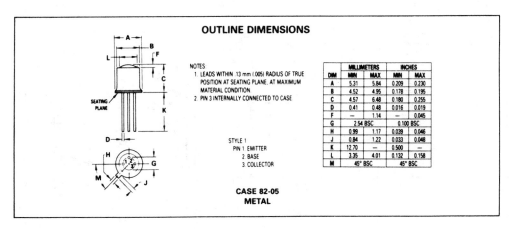

NOTES:
1. LEADS WITHIN .13 mm (.005) RADIUS OF TRUE
   POSITION AT SEATING PLANE, AT MAXIMUM
   MATERIAL CONDITION.
2. PIN 3 INTERNALLY CONNECTED TO CASE.

STYLE 1
PIN 1. EMITTER
    2. BASE
    3. COLLECTOR

CASE 82-05
METAL

DIM	MILLIMETERS		INCHES	
	MIN	MAX	MIN	MAX
A	5.31	5.84	0.209	0.230
B	4.52	4.95	0.178	0.195
C	4.57	6.48	0.180	0.255
D	0.41	0.48	0.016	0.019
F	—	1.14	—	0.045
G	2.54 BSC		0.100 BSC	
H	0.99	1.17	0.039	0.046
J	0.84	1.22	0.033	0.048
K	12.70	—	0.500	—
L	3.35	4.01	0.132	0.158
M	45° BSC		45° BSC	

Figure 7-4   (Continued)

MOTOROLA
SEMICONDUCTORS
P.O. BOX 20912 • PHOENIX, ARIZONA 85036

MFOD71
MFOD72
MFOD73

FIBER OPTIC LOW COST SYSTEM
FLCS DETECTORS

. . . designed for low cost, short distance Fiber Optic Systems using 1000 micron core plastic fiber.

Typical applications include: high isolation interconnects, disposable medical electronics, consumer products, and microprocessor controlled systems such as coin operated machines, copy machines, electronic games, industrial clothes dryers, etc.

- Fast PIN Photodiode: Response Time <5.0 ns
- Standard Phototransistor
- High Sensitivity Photodarlington
- Spectral Response Matched to MFOE71 LED
- Annular Passivated Structure for Stability and Reliability
- FLCS Package
  — Includes Connector
  — Simple Fiber Termination and Connection (Figure 4)
  — Easy Board Mounting
  — Molded Lens for Efficient Coupling
  — Mates with 1000 Micron Core Plastic Fiber (DuPont OE1040, Eska SH4001)

FLCS LINE
FIBER OPTICS
DETECTORS

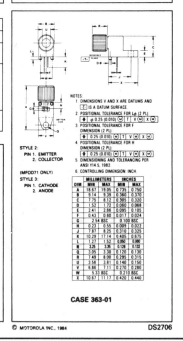

MAXIMUM RATINGS (T_A = 25°C unless otherwise noted)

Rating		Symbol	Value	Unit
Reverse Voltage	MFOD71	V_R	100	Volts
Collector-Emitter Voltage	MFOD72	V_{CEO}	30	Volts
	MFOD73		60	
Total Power Dissipation @ T_A = 25°C		P_D		
MFOD71			100	mW
Derate above 25°C			1.67	mW/°C
MFOD72/73			150	mW
Derate above 25°C			2.5	mW/°C
Operating and Storage Junction Temperature Range		T_J, T_{stg}	−40 to +85	°C

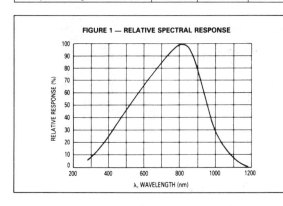

FIGURE 1 — RELATIVE SPECTRAL RESPONSE

λ, WAVELENGTH (nm)

RELATIVE RESPONSE (%)

NOTES:
1. DIMENSIONS V AND X ARE DATUMS AND [T] IS A DATUM SURFACE.
2. POSITIONAL TOLERANCE FOR Lφ (2 PL):
   ⊕ | φ 0.25 (0.010) Ⓜ | T | V Ⓜ | X Ⓜ |
3. POSITIONAL TOLERANCE FOR F DIMENSION (2 PL):
   ⊕ | 0.25 (0.010) Ⓜ | T | V Ⓜ | X Ⓜ |
4. POSITIONAL TOLERANCE FOR H DIMENSION (2 PL):
   ⊕ | 0.25 (0.010) Ⓜ | T | V Ⓜ | X Ⓜ |
5. DIMENSIONING AND TOLERANCING PER ANSI Y14.5, 1982.
6. CONTROLLING DIMENSION: INCH.

STYLE 2:
PIN 1. EMITTER
2. COLLECTOR

(MFOD71 ONLY)
STYLE 3:
PIN 1. CATHODE
2. ANODE

DIM	MILLIMETERS		INCHES	
	MIN	MAX	MIN	MAX
A	18.67	19.05	0.735	0.750
B	9.14	9.39	0.360	0.370
C	7.75	8.12	0.305	0.320
D	1.52	1.72	0.060	0.068
E	2.41	2.66	0.095	0.105
F	0.43	0.60	0.017	0.024
G	2.54 BSC		0.100 BSC	
H	0.23	0.55	0.009	0.022
J	7.87	8.25	0.310	0.325
K	10.29	17.14	0.405	0.675
L	1.27	1.52	0.050	0.060
N	3.25	3.36	0.128	0.132
Q	3.05	3.30	0.120	0.130
R	7.49	8.00	0.295	0.315
U	3.56	3.81	0.140	0.150
V	6.86	7.11	0.270	0.280
W	5.33 BSC		0.210 BSC	
X	10.67	11.17	0.420	0.440

CASE 363-01

© MOTOROLA INC., 1984   DS2706

Figure 7-4   (Continued)

Responsivity:

$$\rho = \frac{I_p}{P_o} = \eta \cdot \frac{e}{h \cdot c} \cdot \lambda \qquad \text{amperes/watt} \qquad (7\text{-}4)$$

Dark current:

$$i_d = 2 \cdot e \cdot I_d \cdot M^2 \cdot F_d \cdot B \qquad \text{amperes} \qquad (7\text{-}5)$$

Shot noise:

$$i_{sn}^2 = 2eiB \qquad (7\text{-}6)$$

Thermal noise:

$$I_{th}^2 = \frac{4K \cdot T \cdot B}{R_L} = i_d B \qquad (7\text{-}7)$$

Total noise:

$$i_n = (i_{sn}^2 + i_{th}^2)^{1/2} \qquad (7\text{-}8)$$

Speed of response:

$$t_d = \frac{x^2}{D} \qquad (7\text{-}9)$$

Cutoff frequency:

$$f_c = \frac{2.8}{2\pi t_d} \qquad (7\text{-}10)$$

Signal-to-noise ratio:

$$\text{SNR} = \left(\frac{\eta \cdot e}{h \cdot f}\right)^2 \cdot \frac{(M \cdot m \cdot P)^2}{i_n} \qquad (7\text{-}11)$$

Detectivity:

$$D^* = \frac{(A \cdot B \cdot R)^{1/2}}{i_n} \qquad (\text{Hz/W})^{1/2} \qquad (7\text{-}12)$$

Noise equivalent power:

$$\text{NEP} = \frac{V_n}{M \cdot \rho \cdot R} \quad \text{W/Hz}^{1/2} = \frac{i_n}{\rho} \qquad (7\text{-}13)$$

QUESTIONS

1. What is one application of an APD?
2. What is responsivity?
3. How are bandgap and wavelength related?

4. What is the major difference between an APD and a PIN?
5. Which type of noise is most significant?
6. What type of transmission uses SNR as a benchmark?
7. Why is quantum efficiency used?
8. What is noise equivalent power?

PROBLEMS

1. If the bandgap energy is 1.35 eV, what is the longest wavelength that the photodiode can detect?
2. The multiplication factor is approximately 100 in a silicon APD, the breakdown voltage is 100 volts, and the material constant is equal to 3. What is the reverse voltage applied?
3. Find the responsivity of the photodetector if the incident power is 20 nW and the photocurrent is 0.3 μA.
4. Find the shot noise if the dark current is 50 nA and the bandwidth is 100 MHz.
5. Using the bandwidth in Problem 4, find the thermal noise if the load resistor is 20 MΩ and the temperature is 298 K.
6. Find the total noise current using Problems 4 and 5.
7. What would the SNR be if the signal power were 17 nW and the noise power were 0.6 nW?
8. Explain how many correct bits would be transmitted with a BER of 10^{-17}.

Modulation Schemes
for Fiber Optics

The student will be able to:

- Distinguish between the different techniques to transmit data.
- Provide terminology necessary to distinguish certain transmitting techniques.
- Understand the difference between modulation and multiplexing.

8.1 INTRODUCTION

Data communication involves the transmission of a signal over some type of channel. *Modulation* is a method of transmitting information on a carrier (light in our case). *Multiplexing* consists of taking many signals and putting them through a single channel. However the signal is transmitted, the object is to make sure that whatever the generated signal looks like, the signal is replicated at the receiving end. Alas, nothing is perfect and the signal ends up being distorted, delayed, or attenuated.

There are modulation schemes to encode a signal so that there can be more than one signal on an individual carrier. The way these encoded schemes are conceived influences the performance of the transmitter and the receiver circuits. Chapters 8 and 9 are closely related. This chapter discusses digital and analog modulation, which is only meant to be a short scenario of the big picture. In the next chapter the circuits necessary to modulate or demodulate the signal will be discussed.

8.2 DIGITAL MODULATION

Digital modulation is sometimes known as a discrete system. Typically, a system is defined in bits per second, bytes per second, or baud rate. Bits per second and megabits per second (Mbps; pronounced "mips") are not the same as baud rate. A *bit* is a logic level 0 or 1. A *byte* is 8 bits. Bit rates are used for parallel transmission where one bit is sent on separate lines. Generally, parallel interfaces are not used in fiber optics.

The baud rate, usually the signaling speed (in seconds), is a true measure of the system's speed. The baud rate is not dependent on data modulation or timing but is defined as the reciprocal of the narrowest pulse width:

$$\text{baud rate} = \frac{1}{\text{narrowest pulse width}} \tag{8-1}$$

$$\text{bandwidth} = BW = \frac{1}{\text{pulse width}} \tag{8-2}$$

$$\text{sampling rate} = (\text{number of channels} \times \text{bit rate}) + 1 \tag{8-3}$$

$$\text{rate} = \text{sampling rate} \times \text{bandwidth} \tag{8-4}$$

The bit rate can be defined as follows:

$$\text{bit rate} = BW \times (\text{samples per cycle}) \times (\text{bits per sample}) \tag{8-5}$$

Example 8-1

A modulation scheme called *pulse code modulation* (PCM) has a 4-byte code and a bandwidth for the analog signal of 4 kHz. If there are two samples per hertz, what is the bit rate?

SOLUTION

$$\text{Bit rate} = 4000 \times (4 \times 8) \times 2 = 256 \text{ kbps}$$

The format in which the signal is to be transmitted influences the data rate. The three most popular digital formats are non-return-to-zero (NRZ), return-to-zero (RZ), and Manchester (Figure 8-1). Other methods, including frequency shift code (FSC) and frequency shift keying (FSK), are also discussed.

8.2.1 Non-Return-to-Zero

Non-return-to-zero (NRZ) is probably the most common form of transmitting a signal. NRZ encoded signals change levels only when the bit level changes. A logic level 1 is represented by a high voltage, approximately 5 V.

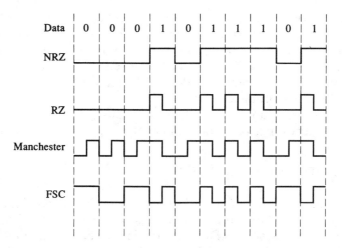

Figure 8-1 Digital modulation schemes.

A logic level 0 is represented by zero volts. For example, if the minimum pulse width is 0.5 μs, the data rate is 2 Mbps. The bandwidth is 2 MHz. Since there is essentially only one voltage level per bit period, the baud rate is 2 Mbaud.

There are problems with this type of transmission. First, it is impossible to distinguish a series of 0's being sent from no signal at all. If a series of 1's are sent, the signal is continuous and coupled electronics will have to be employed for the dc signal. There is also no information about the sampling frequency. A clock synchronized to the sampling rate must be added to interpret the input signal.

8.2.2 Return-to-Zero

The *return-to-zero* (RZ) format is slightly different from NRZ. For a logic level 1, the voltage is high, then low, for each bit period. Logic level 0 keeps a low voltage level for the full bit period. This type of system encoding requires that the system operate twice as fast as the NRZ. Using the example in the NRZ discussion, the bit period is 0.5 μs but the minimum pulse width is 0.25 μs. The bandwidth then becomes 4 MHz and the baud rate is 4 Mbaud. RZ coding must have two intervals per bit since the "1" is high only half of the time of the bit. Another problem with this transmission is that synchronization cannot be found when there is a series of 0's.

8.2.3 Manchester Encoding

Manchester encoding is used for digital data. It is sometimes called *split phase* or *Biphase* modulation. Manchester code requires twice the bandwidth to encode the same bit rate as NRZ. Manchester code represents a logic level 1 as a "1" for the first half of the bit interval and a "0" for the second half of

---

the interval. The case is reversed for a "0" logic level, which is represented as a "0" for the first half of the bit interval and a "1" for the second half. The clock can be recovered from these data by manipulation of the bit stream. If the bit period is still 0.5 μs, the system can be characterized by saying that the minimum pulse width is 0.25 μs. The bit rate is 2 Mbps, with a 4-MHz bandwidth transmitting at 4 Mbaud.

There is no major problem with this coding scheme except for the two-unit interval per bit. This can become a problem at high bandwidths.

8.2.4 Frequency Shift Code

Frequency shift code (FSC) requires the same bandwidth as Manchester encoding. For FSC, the leading edge of each bit requires a transition. The presence or absence of a transition within each bit interval is used to encode a "1" or a "0". For a "1" there is a level change at the center of the bit interval, but for a "0" there is no such change.

8.2.5 Frequency Shift Keying

Frequency shift keying (FSK) has two frequencies keyed to the data, and if the phase of the two frequencies is the same, there is no transition. There are differences between FSC and FSK. FSC requires a transition at the leading edge of each bit, while FSK requires a leading-edge transition only when the two frequencies are out of phase.

8.2.6 Demodulation

The received pulses are demodulated with an *digital-to-analog converter* (DAC). The system is called a *modem* (modulator–demodulator).

8.3 ANALOG MODULATION

Analog modulation is for audio, video, cable TV, broadcast TV, and radar signals at higher bandwidths. The transmitter is turned on and off at the desired rate with digital. In analog, the input is constant over a continuous time frame. The question arises: Why not convert the analog to a digital level? This is not done because of the bandwidth. For example, a standard T1 digital channel requires a much higher bit rate (100 Mbps) and switching speed than that of an analog equivalent at 63 MHz. The circuit will be costlier because of the higher bandwidth. But there has got to be a better way. That way is by using modulation. Modulation is a form of converting a continuous signal to a signal that has amplitude and is sampled at regular intervals.

8.4 MODULATION SCHEMES

In sending any type of data, it can be shown that by transmitting an amplitude $e(t)$ at certain times, no information is lost. This can be accomplished by sampling the signal at certain times, then reconstructing the signal at the receiver end. The value of the signal's amplitude is sampled at at least twice the frequency of the original signal, called the Nyquist rate. These values are sampled in various ways: by amplitude, width, or position.

8.4.1 Pulse Amplitude Modulation

In *pulse amplitude modulation* (PAM) a train of pulses is sent to sample the analog signal. The pulses have a constant width W, which is half the period of the pulse. The amplitude of the pulses corresponds to the amplitude of the signal that is being sent. All the information about the original signal can be found from the height and period of this pulse train, as shown in Figure 8-2.

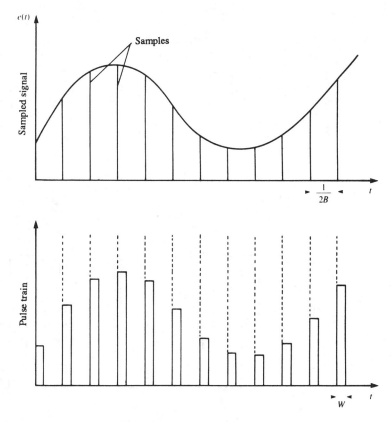

Figure 8-2 Pulse amplitude modulation schemes. (Courtesy of Les Editions Le Griffon D'Argile.)

8.4.2 Pulse Width Modulation

Pulse width modulation (PWM) requires that the pulse train again be sampled at the Nyquist rate. In this instance, the width of the pulse is proportional to the height or amplitude of the original signal. This is shown in Figure 8-3.

8.4.3 Pulse Position Modulation

Pulse position modulation (PPM) is not as easy to implement as the previous types. Again, there is a pulse train, but the pulses are shifted in time by an amount Δt within the interval of the pulse. This time difference is proportional to the amplitude of the original pulse. These pulses have constant amplitude and width. An additional signal for synchronization, such as the clock signal, has to be sent because there would be no information about the frequency of the pulses. PPM is shown in Figure 8-4.

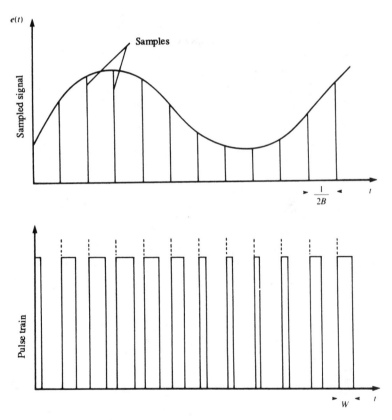

Figure 8-3 Pulse width modulation. (Courtesy of Les Editions Le Griffon D'Argile.)

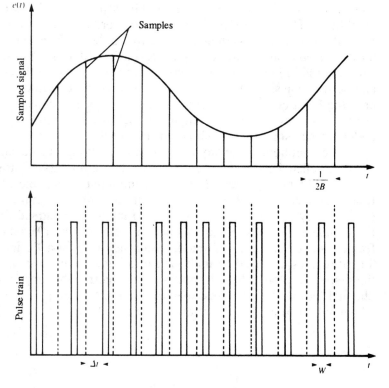

Figure 8-4 Pulse position modulation. (Courtesy of Les Editions Le Griffon D'Argile.)

8.4.4 Demodulation

At the receiver, the train of pulses must be amplified and the process of regenerating the original signal begins. A synopsis of the demodulation schemes is as follows: An ideal passband filter is used in *pulse amplitude demodulation*. Its cutoff frequency is equal to the bandwidth used to filter the pulse trains. The signal can then be recovered. In *pulse width demodulation*, the pulse trains are converted to PAM. This is done by using an integrator and synchronous sampling. The synchronization sampling is recovered by using the rising edge of the PWM signal. For *pulse position demodulation*, PPM is converted to PWM, which as was shown above, is converted to PAM. But it still seems easier to convert the analog signal to a digital signal, and this is done using pulse code modulation.

8.4.5 Pulse Code Modulation

Pulse code modulation (PCM) is a scheme that takes an analog signal and transforms it to a digital signal. The analog signal is sampled at regular intervals, where the amplitude of the sample is based on the current or voltage amplitude. The sample is then quantized or, in effect, rounded off to the nearest level. A code is then assigned to that level. This code is converted to its digital equivalent (for example, level 5 = 101) and transmitted as data. This modulation process is used to reduce the noise level that can occur in the analog transmission of signals of widely varying amplitude. At the receiver end, the PCM codes go through the reverse process (demodulation) to regenerate the analog signal. A sample analog signal and how it is converted to codes for transmission are shown in Figure 8-5.

The telephone company uses PCM to multiplex voice signals. A signaling rate called DS-1 was established to define how 24 separate channels of voice would be digitized. Pulse code modulation samples the signal 8000 times per second, generating 8 bits of information. A frame of 192 bits consists of 1 byte from each channel. A framing bit is added for synchronization—hence 193 bits of information is sent. Since 193 bits are sent 8000 frames per second, the required transmission rate for a T1 line is 1.544 Mbps. T1 rates/lines are used to multiplex into higher transmission rates. The DS lines specify physically (distances) and electrically (impedances) how the channels look. The T rates specify the multiplexing of the DS lines. See Chapter 1 for multiplexing rates for the telephone system.

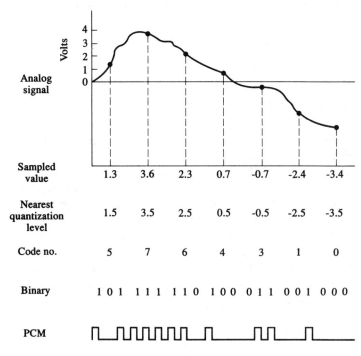

Figure 8-5 Pulse code modulation.

8.5 MULTIPLEXING

The primary reason for using fiber optics is bandwidth. What good is bandwidth if only one signal can be sent? The cost would be prohibitive. The preferred method is to transmit many signals over the same transmission channel. This can be done using frequency, time, or wavelength multiplexing.

8.5.1 Frequency-Division Multiplexing

At this time, modulation of the frequency of light is not possible, but a short discussion on the subject will be useful. *Frequency-division multiplexing* (FDM) transmits several analog signals at the same time over the same channel (Figure 8-6). Each signal is modulated by a different frequency. These modulated signals in turn modulate a subcarrier at a very high frequency. This subcarrier can be modulated in amplitude or frequency. For a fiber optic amplitude modulated system, the modulation changes the radiant flux of the source.

The phone company uses this method to send many signals over one fiber line. Suppose that 4 kHz telephone conversations are being sent. The frequencies that will modulate the signal will be 4 kHz, 8 kHz, . . . , integer × 4 kHz. Therefore, a subcarrier of 2540 kHz can transmit 635 separate telephone conversations. The modulation is processed by bandpass filters to separate the conversations.

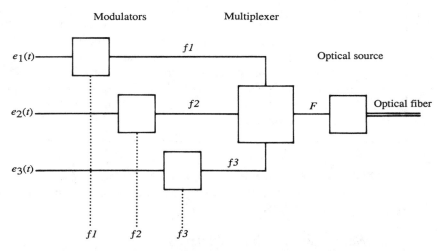

Figure 8-6 Frequency-division multiplexing. (Courtesy of Les Editions Le Griffon D'Argile.)

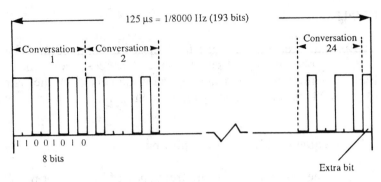

Figure 8-7 Time-division multiplexing. (Courtesy of Les Editions Le Griffon D'Argile.)

8.5.2 Time-Division Multiplexing

Time-division multiplexing (TDM) transmits several digital signals over the same channel (Figure 8-7). Again, to compare multiplexing schemes, a telephone system will be used.

Example 8-2

Find the T1 rate.

SOLUTION

A telephone conversation has a bandwidth of 4 KHz. This time the signal is an 8 bit coded PCM. The sampling frequency is twice the bandwidth, which is 8 kHz or 125 μs. For simplicity say that each bit takes 1 μs, thereby taking 8 μs per second. Subtracting the total bit seconds from the sampling frequency gives 117 μs that the channel is unused. By using the rest of the sampling frequency, 14 other telephone signals can be transmitted. This is the basis for the different T rates in telephony. If 193 bits are transmitted every 125 μs, 24 signals are sent using an 8 bit PCM, which is a T1 rate.

TDM can be used with other pulse modulation schemes (PPM, PAM, PWM). After every twenty-fourth byte is sent, an extra synchronization bit is sent.

8.5.3 Wavelength-Division Multiplexing

Wavelength-division multiplexing (WDM) is specific only to fiber optics. Using several optical sources transmitting at different wavelengths, the light

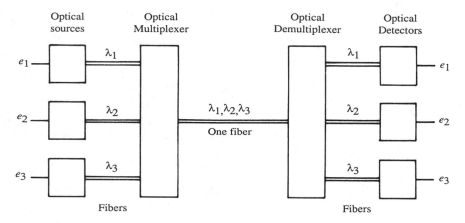

Figure 8-8 Wavelength-division multiplexing. (Courtesy of Les Editions Le Griffon D'Argile.)

is modulated by different electrical signals (Figure 8-8). The different wavelengths are then injected into a single optical fiber. At the receiver end, the light is filtered into separate wavelengths, which are converted into the various electrical signals.

8.6 SUMMARY

There are different ways to transmit information. Digital modulation is the most straightforward. Non-return-to-zero is simply sending the bits as they are; a low level is "0" and a high level is "1." Analog modulation is for video, TV, and higher-bandwidth systems. Different modulation schemes can be used, such as PAM, PWM, PPM, and PCM. Multiplexing takes several signals and puts them through a single channel. Schemes such as time-division multiplexing, frequency-division multiplexing, and wavelength-division multiplexing are used. Wavelength-division multiplexing is used only for fiber optics. The other schemes can be used with coax or fiber. Although this is probably more information than is needed, the reader can begin to see how the encoding schemes and the system definition are related.

8.7 EQUATION SUMMARY

Baud rate:

$$\text{Baud rate} = \frac{1}{\text{narrowest pulse width}} \tag{8-1}$$

Bandwidth:

$$\text{Bandwidth} = \text{BW} = \frac{1}{\text{pulse width}} \tag{8-2}$$

Sampling rate:

$$\text{sampling rate} = (\text{number of channels} \times \text{bit rate}) + 1 \qquad (8\text{-}3)$$

Rate:

$$\text{rate} = \text{sampling rate} \times \text{bandwidth} \qquad (8\text{-}4)$$

Bit rate:

$$\text{Bit rate} = \text{BW} \times (\text{samples per cycle}) \times (\text{bits per sample}) \qquad (8\text{-}5)$$

QUESTIONS

1. Describe amplitude modulation.
2. Describe frequency modulation.
3. What is the difference between PAM and PCM? Which is easier to implement, and why?
4. Explain the Nyquist rate.
5. Describe time-division multiplexing.
6. What is the only multiplexing that can be used by fiber optics exclusively?
7. What is frequency-division multiplexing?

PROBLEMS

1. Derive the T2 rate. Find the bandwidth, sampling rate, and rate. Note that due to overhead bits, the answer may not quite be what the rate is defined as.
2. A Manchester code has a bit period of 0.25 μs, a minimum pulse width of half the bit period, and a bit rate of 2 MHz. What are the bandwidth and the baud rate?
3. Suppose that a NRZ code is being sent with the pulse width of 10 ns. What are the bandwidth and the baud rate?
4. Having the same information as in Problem 3, what are the bandwidth and bit rate for RZ?

9

Practical Optical Transmitters
and Receivers

The student will be able to:

- Investigate the different configurations for transmitters and receivers.
- Distinguish between the different circuits and the modulation schemes.
- Understand how components (resistors and capacitors) can solve some noise problems.

9.1 INTRODUCTION

In previous chapters, optical sources, detectors, and modulation schemes were discussed. This chapter discusses the realization of practical fiber optic transmitters and receivers.

9.2 OPTICAL TRANSMITTERS

Optical transmitter design is determined by the category of the optical source (LED or ILD) and by the type of transmission (analog or digital). A generalized schematic is shown in Figure 9-1. The input signal can be digital or analog. The electronic preprocessing takes the input voltage and converts it to current to modulate the LED or laser. The standard interface circuit converts the input voltages and impedances to electrical signals for the drive

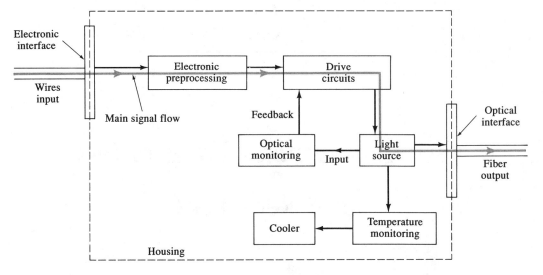

Figure 9-1 Typical transmitter circuit. (Courtesy of Sams: A Division of Macmillan Computer Publishing.

circuit. The drive circuit, using a type of transistor network, maintains the quiescent voltages that operate the LED or laser. Optical monitoring is used to stabilize the lasers. A small amount of light is emitted from the rear face of the laser, which is sensed by the photodetector. Changes in the transmitter output triggers a feedback circuit to adjust the drive current. Temperature monitoring assures operation at stable operating characteristics. By keeping a constant temperature, the threshold current, output power, and wavelength will not change. The following sections discuss each combination of transmitter category and transmission type for analog or digital independently.

9.2.1 LED Transmitter Design: Digital

One design requirement for a digital transmitter is that the risetime be small enough for the specific bit rate. The risetime requirement for a typical transmitter ranges from 1 to 20 ns, with a nominal value of 10 ns, to achieve a bit rate of several hundred Mbps. The LED is easier to operate and can be switched on and off at high speeds in response to the drive circuitry, producing current as needed.

Figure 9-2 uses a bipolar transistor (such as a 2N2222 or 2N3904) in a common emitter mode. This single stage circuit provides current gain for the LED. When the transmitter is in saturation (collector voltage is forward biased), the emitter to collector voltage, V_{CE}, is around 0.3 V; hence the LED is off. The maximum current flow through the LED is limited by the value of R_2. When independent bias is wanted, R_3 is added. The common emitter circuit is limited in switching speed because of the diffusion capacitance and the space charge. This circuit can transmit up to 30 MHz.

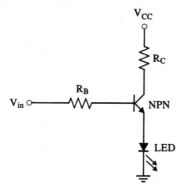

Figure 9-2 Common-emitter digital drive circuit.

The switching speed can be increased by using a low-impedance drive circuit which rapidly charges up the capacitance. This can be done with an emitter follower drive circuit. These designs use a low impedance current source, such as the collector of a bipolar junction transistor, to drive the LED. The drive circuit must be capable of driving from 100 to 200 mA with 1.4 to 1.6 V biasing, as shown in Figure 9-3. The circuit uses an emitter-follower configuration. The compensating matching network (R_3C) modulates the LED directly. The capacitance of 180 pF would allow for 100 Mbps operation of the LED.

For digital transmission below 10 MHz NRZ, common TTL logic gates are used to interface to the LED. Line drivers can be used to achieve direct compatibility with TTL logic, as shown in Figure 9-4a. Texas Instrument's 75452 provides a drive current of about 60 mA to the LED when R_L is 50 Ω. The value of R_1 is a function of the power supply voltage and required LED drive current. The risetime is about 5 ns.

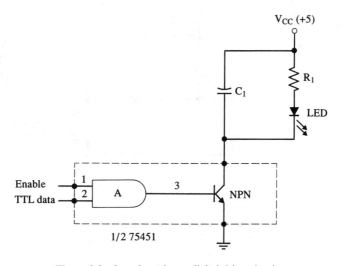

Figure 9-3 Low-impedance digital drive circuit.

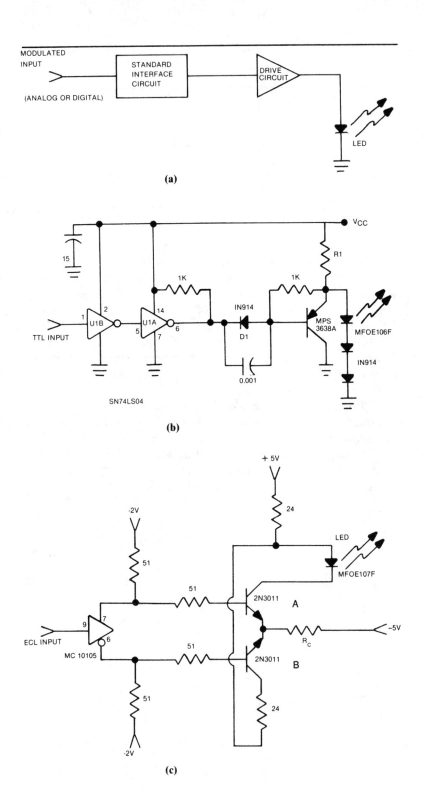

(a)

(b)

(c)

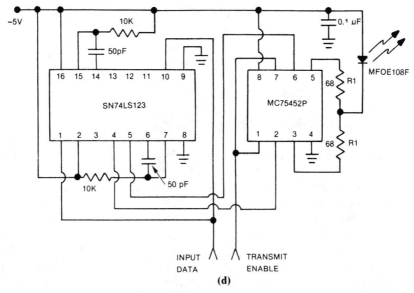

Figure 9-4 (a) TTL LED drive circuit; (b) TTL shunt drive circuit; (c) emitted-coupled logic circuit; (d) Manchester transmitter. (Courtesy of AMP, Inc.)

A shunt configuration can give better risetimes for the LED and is shown in Figure 9-4b. The SN74LS04 is used to accept TTL signals and provide the necessary drive current. When the transistor is switched on, the LED is off, and vice versa. With the drive current passing either through the LED or the transistor, ripple from the power supply can be alleviated. If the transistor is in saturation, the diodes in series with the LED ensure that the LED is off. Diode D_1 prevents reverse-bias voltage damage. The desired LED drive current, I_f, is limited by the resistor R_L. To derive the drive current the following equation is used, where V_{CC} is the power supply voltage:

$$I_f = \frac{V_{CC} - 3 \text{ V}}{R_L} \tag{9-1}$$

If the bandwidth needs to be still higher, other logic families can be used, such as *emitter coupled logic* (ECL). The LED acts as a load in one collector, which provides current gain and drive current for the device. This circuit resembles a linear differential amplifier in the switching mode. The nonsaturating characteristics of this configuration give fast switching times because the charge is stored at the transistor base. The MC10105 is a driver for the circuit, providing the voltage levels for the transistors. The transistors labeled A and B in Figure 9-4c are used to provide the necessary ECL voltage levels. A high level is roughly -0.8 V and a low about -1.8 V when the anode is at zero volts. This particular circuit has a 50 Mbps bandwidth. With faster switching transistors and ECL logic, this circuit can be used up to 300 Mbps.

The transmitter types described above use NRZ data with a 50% duty cycle. Other data schemes, such as 2 Mbps Manchester transmitter, can be implemented as shown in Figure 9-4d. Manchester transmission requires a constant optical output. Positive and negative transitions from the quiescent level are used to form the digital information. A driver (MC75452P) provides the positive and negative drive current. The SN74LS123 is a monostable multivibrator that pulses the driver. This circuit can be extended to 10 Mbps if the risetime of the components and the LED are chosen accordingly.

9.2.2 LED Transmitter Design: Analog

For analog transmission the drive circuitry must follow a constantly varying signal in amplitude and phase. The LED has the inherent quality of being more linear than ILDs. When analog modulation is used, the LEDs exhibit nonlinearities that cause intermodulation distortion among multiple channels. Nevertheless, there are some analog applications such as a single video channel over fiber in which analog circuits are needed.

A simple analog modulation scheme is shown in Figure 9-5a. It consists of a Class A amplifier using a 2N3904 transistor. The transistor collector current, i_c, is the LED's drive current. V_{dc} together with R_a and R_b provide the dc base current, I_b. I_b forward-biases the base–emitter junction, turning the transistor on. The Q point is chosen such that the base current does not turn the transistor off during the negative swing of the analog signal. During the positive swing, the transistor does not go into saturation. R_e stabilizes the operating point. The collector current is always well above cutoff. Another analog design, shown in Figure 9-5b, uses a Darlington pair to reduce the impedance of the source.

9.2.3 ILD Transmitter Design: Digital and Analog

The design of a digital ILD transmitter is about the same as that for the LED. The drive current is different (100 to 200 mA), with voltages of 1.6 to 1.8 V.

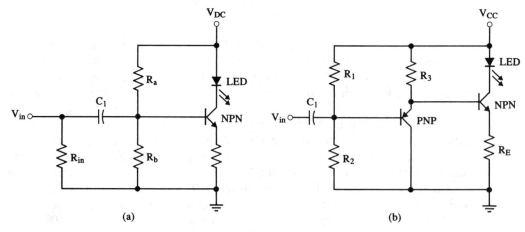

Figure 9-5 (a) Analog LED design; (b) low-impedance analog LED.

ILD devices have the advantage of not needing compensation circuitry because they have a wider bandwidth. An ILD drive circuit (Figure 9-6a) uses a shunt driver with a field effect transistor to provide for high-speed operation. R_2 and C are used to bias the field effect transistor into the active pinch-off region. This type of configuration could modulate the ILD to greater than a 1-gigabit bandwidth.

Other high-speed digital modulation circuits can be used. In Figure 9-6b, two differential amplifiers are connected in parallel. The input voltage stage

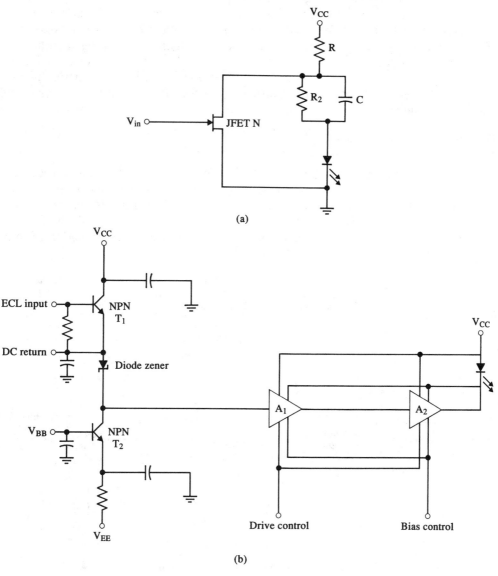

(a)

(b)

Figure 9-6 (a) Digital ILD shunt driver; (b) ECL digital ILD.

uses an emitter follower T_1. T_2 acts as a current source with a zener diode to adjust the level for the emitter-coupled logic. The differential amplifiers provide sufficient modulation current for the laser. ILDs are rarely employed for analog transmission because of their nonlinearities.

9.3 OPTICAL RECEIVERS

The optical receiver must convert optical energy into usable electrical signals. Figure 9-7 shows a generalized optical receiver circuit. Notice that the analog and digital are similar, at least for the front end. Even though the signal was sent digitally, by the time it reaches the receiver it varies continuously, like an analog signal. The decision circuit after the analog amplification converts it back to digital form. The output of the detector is converted in the amplifier to a higher current. The processing block is where demodulation occurs. The last block provides the rest of the circuit with the proper signal levels and format.

9.3.1 Receiver Sensitivity

Receiver sensitivity specifies the weakest optical signal that can be received. The minimum signal that can be detected depends on the noise of the receiver's front end. The noise of the amplifier is included in the design calculation (see Chapter 7). After the noise floor limit has been established, the SNR or BER of the receiver must be chosen. Sensitivity can be expressed in dBm or

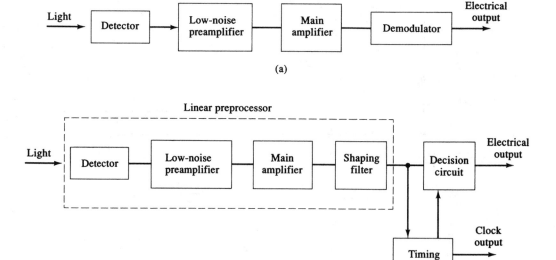

Figure 9-7 Typical receiver circuits: (a) analog; (b) digital. (Courtesy of Sams: A Division of Macmillan Computer Publishing.)

microwatts. In the specifications of most packaged receivers, the sensitivity is always given with respect to some level of performance, such as BER.

9.3.2 Dynamic Range

Dynamic range is the difference in the minimum and maximum acceptable power levels. The minimum is set by the detector's sensitivity. The maximum is set by the amplifier of the detector. Power levels above the maximum distort the signal and saturate the receiver. If the dynamic range is exceeded in an analog system, such as an analog speaker, the speaker is driven with more power than it can handle. In a digital system, exceeding the dynamic range can increase the BER. If a receiver has a minimum optical requirement of −30 dBm and a maximum of −10 dBm, its dynamic range is −20 dBm. Optical received inputs are therefore in the 1 to 100 μW range.

9.3.3 Amplifier

There are two basic designs for the amplifier portion of a receiver: low impedance and transimpedance (Figure 9-8). The bandwidth of the amplifier is determined by the *RC* time constant of the circuit. The gain has to be used for the transimpedance amplifier. The basic formula is as follows:

$$BW = \frac{\text{gain}}{2\pi RC} \qquad (9\text{-}2)$$

The gain, or more exactly the closed-loop gain, for the low-impedance amplifier is 1.0. For the transimpedance amplifier gain values range from 10 to 1000.

9.3.4 Duty Cycle of the Receiver

Some optical receiver designs put restrictions on the duty cycle. A duty cycle far below 50% will cause the receiver to misinterpret the high or low signal

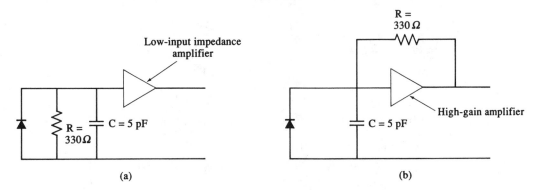

Figure 9-8 Amplifier receivers: (a) low-impedance; (b) transimpedance.

level significantly. The high or low signal level is distinguished by a reference threshold level set by the amplifier. If the duty cycle drifts from the ideal 50% level, the BER will be much lower.

There are trade-offs, as always, between the design and the duty cycle. Manchester code could be used, since by definition it has to have a 50% duty cycle. But as has been seen, twice the bandwidth is needed and an increase in the circuit complexity is needed for the receiver. The other solution is to design a receiver that maintains threshold without drift. This type of receiver would only be good for digital pulses. If analog is to be used, dc levels can be removed from the signal capacitively. A capacitor could be placed between the photodetector and the amplifier.

9.3.5 Comparison of Receiver Design

The sensitivity of the receiver depends on the modulation method employed. Once the modulation method is known, the signal-to-noise ratio (SNR) and expected bit error rate (BER) can be calculated. This section will compare two types of practical receivers: a low-impedance receiver and a transimpedance receiver. They both incorporate the PIN photodiode. To make the comparisons fair, the bandwidths of both receivers is set at 100 MHz and a non-return-to-zero (NRZ) modulation scheme is used.

9.3.5.1 Low-impedance receiver. The low-impedance amplifier bandwidth is determined by the RC input network shown in Figure 9-8a. The closed-loop voltage gain of the amplifier cell is assumed to be 30. For maximum sensitivity, the input resistance of.the amplifier is assumed to be infinite. Since a PIN photodiode is being used, the responsivity is 0.4 mA/mW, the multiplication is 1.0, and the excess noise factor is also 1.0. To further demonstrate the noise performance of the receiver, the 1's and 0's received will be evenly distributed. The assumed levels for various receiver parameters are as follows:

$$P_{opt} \text{ (``1'' level)} = P_1 = 1 \ \mu W = -30 \ dBm$$

$$P_{opt} \text{ (``0'' level)} \quad = 0 \ \mu W$$

$$P_{average} \quad = P_a = 0.5 \ \mu W = -33 \ dBm$$

$$I_{dark} \quad = 5 \ nA$$

$$T \quad = 300° \ Kelvin$$

To make the equations more meaningful, a list of the abbreviations is given.

B_s	signal bandwidth
C_{eq}	equivalent capacitance of the preamplifier
g	gain of the amplifier
I_{dark}	dark current

i_s shunt current

K Boltzmann's constant

e charge on an electron

η quantum efficiency

ρ responsivity

The SNR now can be calculated for a "real" receiver as shown in Example 9-1.

Example 9-1

Calculate signal bandwidth, NEP, and SNR for the low-impedance amplifier (Figure 9-8a).

$$\text{Signal bandwidth} = B_s = \frac{1}{2\pi \cdot R_{eq} \cdot C_{eq}} = 96.46 \text{ MHz} \qquad (9\text{-}3)$$

"1"-Level output

$$= v_1 = -R_{eq} \cdot g \cdot i_s = -R_{eq} \cdot g \cdot \rho \cdot P_1 = -4 \text{ mV} \qquad (9\text{-}4)$$

$$\text{Noise equivalent bandwidth} = B_n = \frac{1}{4 \cdot R_{eq} \cdot C_{eq}} = 150 \text{ MHz} \qquad (9\text{-}5)$$

Average output noise $= v_n^2$

$$= 2 \cdot g^2 \cdot R_{eq}^2 \cdot B_n \cdot \left(e \cdot \rho \cdot P_a + e \cdot I_{dark} + \frac{2 \cdot K \cdot T}{R_{eq}} \right) \qquad (9\text{-}6)$$

$$= 7.4 \times 10^{-7} \text{ V}^2$$

$$\text{Noise equivalent power} = \text{NEP} = \frac{v_n}{g \cdot \rho \cdot R_{eq}}$$

$$= 216 \text{ nW (average)} = -36.6 \text{ dBm}$$
$$(9\text{-}7)$$

$$\text{Signal-to-noise ratio} = \text{SNR} = 10 \log \left(\frac{v_1^2}{v_n^2} \right)$$
$$(9\text{-}8)$$
$$= 13.4 \text{ dB (electrical)}$$

For these calculations resistor noise is the most dominating factor. If the desired bit error rate (the number of false bits received), is to be less than 10^{-9}, an SNR of 13.4 dB should be sufficient.

9.3.5.2 Transimpedance amplifier.   Figure 9-8b shows the transimpedance amplifier using a PIN photodiode. The bandwidth is determined by the feedback resistor and the capacitance, divided by the open-loop gain of the amplifier. Assuming the same gain, "1" and "0" optical levels, average power, dark current, and temperature as used for the low-impedance amplifier, the equations in Example 9-2 apply.

Sec. 9.3   Optical Receivers   **157**

Example 9-2

Calculate signal bandwidth, NEP, and SNR for the transimpedance amplifier.

$$\text{Signal bandwidth} = B_s = \frac{g}{2\pi \cdot R_{eq} \cdot C_{eq}} = 96.46 \text{ MHz}$$

$$\text{``1''-level output} = v_1 = -R_{eq} \cdot i_s = -R_{eq} \cdot \rho \cdot P_1 = -4 \text{ mV}$$

$$\text{Noise equivalent bandwidth} = B_n = \frac{g}{4 \cdot R_{eq} \cdot C_{eq}} = 150 \text{ MHz}$$

$$\text{Average output noise} = v_n^2$$

$$= 2 \cdot R_{eq}^2 \cdot B_n \cdot \left(e \cdot \rho \cdot P_a + e \cdot I_{dark} + \frac{2 \cdot K \cdot T}{R_{eq}} \right)$$

$$= 8.18 \times 10^{-12} \text{ V}^2$$

$$\text{Noise equivalent power} = \text{NEP} = \frac{v_n}{g \cdot \rho \cdot R_{eq}}$$

$$= 13.9 \text{ } \mu\text{W (average)} = -18.57 \text{ dBm}$$

$$\text{Signal-to-noise ratio} = \text{SNR} = 10 \log \left(\frac{v_1^2}{v_n^2} \right)$$

$$= 170.37 \text{ dB (electrical)}$$

This SNR is better than that for the low-impedance amplifier. This design is more stable and linear, and not as dependent on the amplifier gain as is the low-impedance type.

9.3.6 Other Amplifiers

Other types of amplifiers using field effect transistors and bipolar junction transistors are very popular. The FET's dominant noise source, called *channel noise*, is equal to the total capacitance divided by the square root of the gain. The bipolar amplifier's primary noise contributions are the base current shot noise and collector current shot noise. No matter what type of preamplifier is used, the total signal-to-noise ratio must be calculated or known in order to design a fiber optic system.

9.4 TRANSCEIVERS AND REPEATERS

In some applications, such as a remote computer terminal to a mainframe, it is desirable to have a transmitter and receiver in a single package. This is called a *transceiver*. The transceiver sends and receives a signal usually over two separate fiber cables. The dual circuits are isolated from one another.

A repeater contains a receiver and a transmitter but is connected in series. The receiver detects the signal, amplifies and regenerates it, and produces an electrical signal that drives the transmitter in the repeater. Repeaters are used in long-span links such as telecommunications.

9.5 PACKAGING

There are two major packaging options for the optical source or receiver. Usually, transistor outline cans (TO18/TO52) are used. Another packaging option is a pigtailed source, frequently selected when the designer wants maximum coupling into the fiber, as shown in Figure 9-9. Detectors can be prepackaged to include automatic gain control, demultiplexing, and amplifier all in a single DIP package.

9.6 SUMMARY

Optical transmitters are designed for the particular application such as digital or analog. From here the designs are as complex or as simple as the designer wants to make them. An optical receiver must be able to efficiently change the optical energy into electrical current. Preamplifiers and amplifiers needed to boost the signal make receivers the noisiest part of the system. The design must be correct to keep the signal above the noise.

9.7 EQUATION SUMMARY

LED drive current for a shunt configuration:

$$I_f = \frac{V_{cc} - 3\text{ V}}{R_L} \tag{9-1}$$

Bandwidth of a receiver amplifier:

$$BW = \frac{\text{gain}}{2\pi RC} \tag{9-2}$$

Signal bandwidth for a receiver:

$$\text{signal bandwidth} = B_s = \frac{1}{2\pi \cdot R_{eq} \cdot C_{eq}} \tag{9-3}$$

Logic-level output for a receiver:

$$\text{``1''-level output} = v_1 = -R_{eq} \cdot g \cdot i_s = -R_{eq} \cdot g \cdot \rho \cdot P_1 \tag{9-4}$$

Noise equivalent bandwidth for a receiver:

$$\text{Noise equivalent bandwidth} = B_n = \frac{1}{4 \cdot R_{eq} \cdot C_{eq}} \tag{9-5}$$

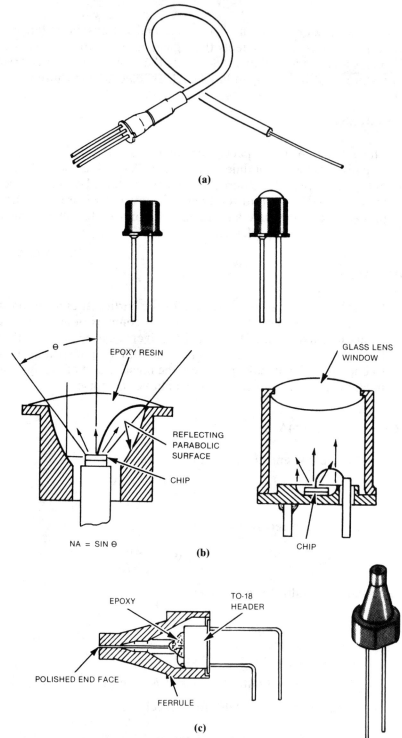

(a)

EPOXY RESIN

Θ

REFLECTING
PARABOLIC
SURFACE

CHIP

NA = SIN Θ

GLASS LENS
WINDOW

CHIP

(b)

EPOXY

TO-18
HEADER

POLISHED END FACE

FERRULE

(c)

Figure 9-9 Packaging of transmitters and receivers: (a) pigtail; (b) TO-style packages; (c) ferruled source. (Courtesy of AMP, Inc.)

Average output noise for a receiver:

Average output noise

$$= v_n^2 = 2 \cdot g^2 \cdot R_{eq}^2 \cdot B_n \cdot \left(e \cdot \rho \cdot P_a + e \cdot I_{dark} + \frac{2 \cdot K \cdot T}{R_{eq}} \right) \qquad (9\text{-}6)$$

Noise equivalent power for a receiver:

$$\text{Noise equivalent power} = \text{NEP} = \frac{v_n}{g \cdot \rho \cdot R_{eq}} \qquad (9\text{-}7)$$

Signal-to-Noise ratio of a receiver:

$$\text{Signal-to-noise ratio} = \text{SNR} = 10 \log \left(\frac{v_1^2}{v_n^2} \right) \qquad (9\text{-}8)$$

QUESTIONS

1. An ILD is used for which type of transmission, digital or analog?
2. In your own words, explain a receiver circuit.
3. What are the basic elements necessary for a transmitter circuit?
4. What is the major drawback of any receiver circuit?
5. What is the difference between a repeater and a transceiver?

PROBLEMS

1. A receiver has a sensitivity of -30 dBm and a maximum receivable power of -15 dBm. What is the dynamic range?
2. If a receiver has a sensitivity of -27 dBm, what is the minimum optical power that it can receive?
3. A transimpedance amplifier's receiver has a closed-loop gain of 10. The equivalent resistance is 150 Ω and the capacitance is 40 μF. What is the bandwidth?
4. Using the average values for the digital transmitter shunt configuration, design a transmitter circuit showing the biasing.

10

Systems Architecture

CHAPTER OBJECTIVES

The student will be able to:

- Distinguish between various networking topologies.
- Briefly describe each type of networking scheme.
- Describe briefly the different standards for networks.

10.1 NETWORKS

Digital networks have evolved since 1962, when AT&T realized that analog circuits were expensive to maintain and design. Communication networks today integrate voice, video, and data over a common transmission medium. With the ever-increasing need for higher bandwidths, fiber optics is the medium of choice. Networks are divided into different schemes: point-to-point links, local area networks, metropolitan area networks, and wide-area networks.

10.2 POINT-TO-POINT LINKS

Point-to-point links are the simplest of the networks, starting at one point and going to another point. The major problem with this type of link is that when the cable is cut, at least part or all of the system link will go down. This would be a catastrophic failure for most communications networks, so other designs are needed.

10.3 LOCAL AREA NETWORK

A *local area network* (LAN) is an electronic communication network limited to a geographical area such as a large office or a building. LANs are limited in size to less than 1 km. Communication links connect personal computers, mainframes, printers, plotters, and facsimile machines. Shared services, databases, electronic mail, application software, and other services can all be maintained on this network. The network incorporates a transmission medium that connects several nodes or stations. At each node, electronic equipment, run by the software and hardware, is connected to the network.

The network has different configurations called *topologies* that define the physical and logical arrangements. Figure 10-1 shows some examples of the principal topologies: star, ring, and bus. *Star* LANs have a central hub or controller in which the nodes radiate out from that point. The central hub acts like a controller, routing signals from one node to another. The hub can use a passive star coupler or an active controller. The *ring* network links all terminals in a point-to-point series. Messages flow in one direction only from node to node. If one node fails, the system is down unless bypass components are used. The *bus* network has each node tapped off from the transmission medium. Messages flow in either direction along the bus.

Any configuration of one or all these topologies are used. For example, a system may be a physical star/logical ring topology. This means that the physical fiber optic portion of the link looks like a star. The logical arrangement of the hardware and the software looks like the ring configuration.

A scheme known as *multiple access* is used for LANs. The transmission medium and the bandwidth are shared among the nodes. Multiple-access

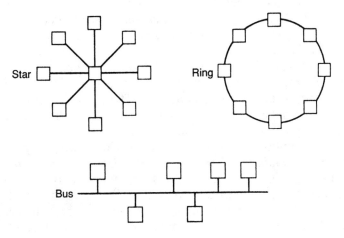

Figure 10-1   Network topologies.

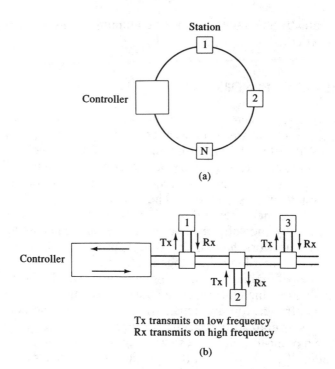

Figure 10-2 Dedicated access schemes: (a) time-division multiple access; (b) frequency-division multiple access.

schemes are classified as dedicated or contention. *Dedicated systems* have a portion of the communications channel allocated to each station or node. *Contention systems* bid for the use of the channel. There are several approaches to multiple-access schemes.

Dedicated access is divided into two groups: *time-division multiple access* (TDMA) and *frequency-division multiple access* (FDMA). These are shown in Figure 10-2. TDMA uses a controller on the LAN to set up transmission frames that contain time slots. Each time slot or slots is dedicated to each station desiring access. FDMA divides the access into frequency bands, one assigned to each station wanting access. The node transmits on one frequency and receives on another. The cable TV industry uses this method to provide data services. A translator is used to distinguish between transmitting and receiving frequencies.

Contention access methods are divided into groups, *carrier-sense multiple access/collision detection* (CSMA/CD) and *token passing*, as shown in Figure 10-3. Contention methods are more widely used because of the more intelligent use of the transmission medium and the bandwidth. When using CSMA/CD, each node listens, sensing the carrier, before it attempts to access the channel. The node transmits only when the channel is clear. If two or more stations transmit at the same time, a collision occurs. A collision signal

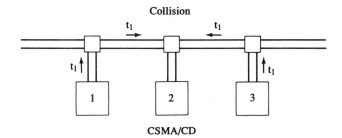

Figure 10-3 Contention access schemes.

is sent to all nodes, causing all transmissions to cease. After a predetermined time frame, the nodes try to transmit again.

Token passing polls the nodes, giving them the opportunity to transmit on the channel, as shown in Figure 10-4. A control message, called a *token*, gives the right to access the channel. The token is passed from one node to another. Only the station with the token can transmit. If there is nothing to transmit, the token is regenerated and passed on to the next station. The token ring operates at two rates, 4 and 16 Mbps.

There are also self-healing LANs: If one node goes down, the controller can reconfigure itself and bypass that node. Alarms are sent to another site so that the node that is down can be replaced or repaired. Self-healing networks will provide central office backup and alternate routing for customer traffic.

10.4 METROPOLITAN AREA NETWORKS

Metropolitan area networks (MANs) transport signals within a city, between buildings, and even between other LANs. MANs are used for telephone local loops, DS-1 and DS-3 business networks, cable TV, and private business networks. LANs are point-to-point links, leading to spider web–like designs. Complex and costly arrangements of multiplexers with add-drop facilities are needed. A backbone or major network in which LANs are interconnected on a demand basis is used instead. A MAN can provide speed and distance capability over a broader area. MANs can also be used instead of costly dedicated leased lines.

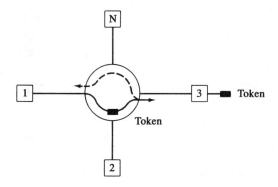

Figure 10-4 Token-passing ring.

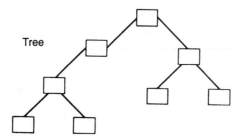

Figure 10-5 Tree topology.

MANs employ the same type of topology as that used for LANs, with one additional configuration available. The MAN can carry voice, video, and data information while linking the information into LANs. This requires a tree topology of the type shown in Figure 10-5. Although the topologies look the same, the differences come in the form of how the medium and channel capacity are shared. With a LAN, the channel and the medium are shared equally. With a MAN, redundancy is built into the system, and alternate routes are available. Since a metropolitan area has a high concentration of construction, traffic, and exposure to cable damage, route diversity is critical. In most topologies, the signal is routed in the opposite direction, usually using another route.

Applications for MANs are campus networks (such as universities), industries, and research laboratories that need to connect the LANs. The user will be able to access large databases, file servers, and voice and other fixed bandwidth traffic from one site to another. An example of a MAN is shown in Figure 10-6. The MAN uses a slotted structure for the packets. The time slots are segmented, accommodating 48 bytes of data plus a header. A 44-byte section is left for the user's data.

10.5 WIDE-AREA NETWORKS

Wide-area networks (WANs) are used in long-haul trunking between two points. The telephone industry uses WANs between central offices and switching centers. Fiber optics is used for WANs because of the channel

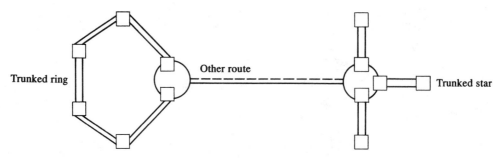

Figure 10-6 MAN cluster.

Systems Architecture   Chap. 10

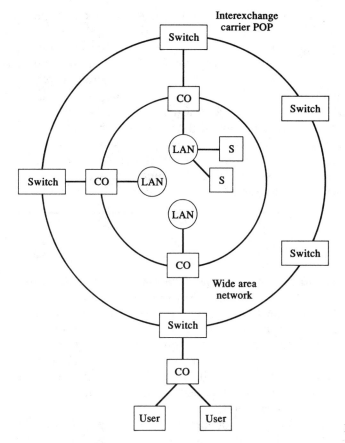

Figure 10-7 Wide-area network.

capacity, ranging from 560 Mbps to 2.4 Gbps. Figure 10-7 shows a WAN configuration. In Figure 10-7, WANs are connected to MANs which are connected to LANs. This can get confusing, but it will be easy to see the need for all these networks.

Example 10-1

In this example, the linking of a LAN, MAN, and WAN is illustrated. The user calls a customer in another city in the state. This call is routed from the user on a LAN. The voice, which has been converted digitally, is sent through a fiber optic MAN to the central office (CO). At the CO the person's voice, or data, is received in a switch or digital cross connect (DACS). The signal is then converted back to electricity. At the DAC, the signal is multiplexed into higher speed groups, generally T3 rates. The signal is passed to an interexchange carrier point of presence (POP). At the POP, the channels enter the interexchange carriers switch or DACS, to be routed to the long-haul network switch within the net-

---

work. The user's voice, which needs to be put on the network, will be multiplexed at a T3 rate. The voice signal is connected through a protection switch to the fiber optic terminal equipment. The fiber optic terminal equipment contains higher speed multiplexing that combines the channels further, to an upward limit of 2.4 Gbps. The protection switch is used to protect against outages between the terminal and the repeater. An overhead signal, orderwire (acknowledgment from the repeater), alarm, error encoding, and control data between the POPs or repeater sites are encoded for transmission. Finally, the signal is again converted to light and sent on a fiber optic WAN. The customer finally gets the call, after the entire process has been repeated backwards at the other end. It is amazing that there is little delay in this entire process.

10.6 STANDARDS

Many networks are controlled by standards specifically designed for both the electrical and optical portions of the network. Previous to standards committees being formed, the fiber industry relied on de-facto standards and word of mouth to choose connectors, transmission equipment, and installation practices. The breakup of AT&T and infusion of foreign suppliers led to a lot of confusion and mismatch of parts and components. That was in the 1980s, the decade of confusion in the fiber industry. In the 1990s, the fiber industry and the standards will meet. Measurements and performance of the fiber, the system, and the network will be well defined.

The first standards to emerge were for measurements of physical cable parameters. The military first defined how to measure the mode field diameter, attenuation, and tensile strength. DOD Standard 1678 was the first widely used measurement standard. Organizations such as the National Bureau of Standards (NBS), the Electronic Industries Association (EIA), the American National Standards Institute (ANSI), Consultative Committee in International Telegraphy and Telephony (CCITT, Geneva), and the Institute of Electrical and Electronics Engineers (IEEE) are just a few of the organizations involved in standardization. Appendix B lists major standardization organizations and the standards important to fiber optics, digital, or network design.

10.7 ETHERNET

The *Ethernet* network standard was developed by Xerox. A host computer or workstation is attached to a coaxial cable (Figure 10-8). The transmission line operates at 3 Mbps. Data are transmitted in the form of a packet, constituting up to 4000 data bits. Addressed packets let the station know when a packet is addressed to it. The station can then remove the packet from the line and "read" it. The workstations can detect when the line is free and transmit at any time to another station. It is also possible for two workstations to trans-

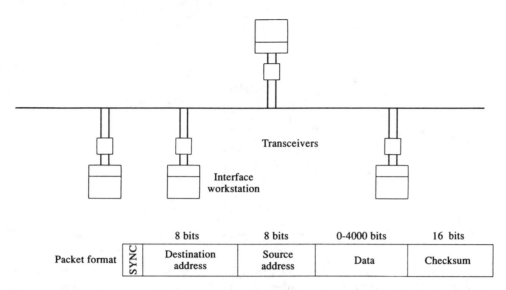

Packet format	SYNC	Destination address	Source address	Data	Checksum
		8 bits	8 bits	0-4000 bits	16 bits

Figure 10-8 Ethernet network topology and packet format.

mit at the same time. When a collision occurs, transmission ceases, only to resume after a given amount of time. This is known as *statistical multiplexing*, or CSMA/CD protocol, described earlier.

As noted above, Ethernet calls for coax as the medium. Fibernet, also from Xerox, is the fiber optic version of Ethernet. Two versions emerged for the Fibernet CSMA/CD protocol. One version, called the 10 Base-FA, uses an active star coupler in the point-to-point link and allows for 2 km interstations. The second version, 10 Base-FP, uses a passive star and allows only a 1.1 km interstation. In the second version, 19 passive transmissive star couplers were used, with only 10 dB loss between two ports. Data can now be transmitted at higher rates, 150 Mbps over distances of 0.5 km, with zero errors.

10.8 IEEE 802.6

IEEE 802.6 is a part of several standards for local and metropolitan area networks. It defines a *distributed queue dual bus* (DQDB) protocol. The other standards include 802.3, which describes CSMA/CD; 802.4, token bus; 802.5, token ring; and 802.9, integrated voice and data LAN. What is unique about this standard is that it defines a dual-bus subnetwork with two unidirectional, oppositely directed physical buses. Figure 10-9 shows such a design. The buses, designated A and B, generate fixed-length slots that carry segmented time slots from node to node. The nodes reserve empty slots by sending requests on B and capturing the bus on A. A counter is used at each station to count the empty slots already reserved on the bus. The station can then use an empty slot.

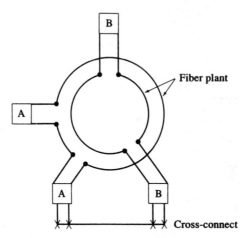

Fiber plant

Cross-connect   **Figure 10-9** Dual-bus subnetwork.

This standard also describes bridging the network, to form a MAN, by using routers and bridges. Routers are used to handle large network requirements. The routers, which are protocol dependent, gather information about the network continuously, sharing it throughout the network. Flow control and routing are updated constantly so that the network does not get in a bottleneck. Source routing bridges are used with end stations that request the bridge to flood the network with route discovery frames. The returned frames are used by each station to determine a specific route. The stations store this information, updating the route table periodically.

10.9 ISDN

ISDN is an acronym for *integrated-services digital network*, an international standard. The Consultative Committee in International Telegraphy and Telephony (CCITT) is responsible for international standards. The primary purpose behind ISDN is the integration of new services with existing services. ISDN expands the existing facilities by providing two voice channels (2B) and

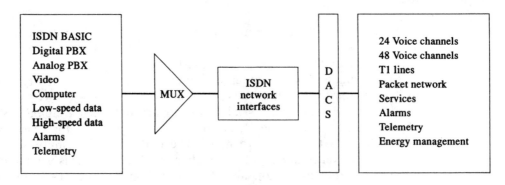

Figure 10-10 ISDN digital network.

a data channel (D) on a single subscriber pair. Broadband ISDN is predicted to be the choice for fiber to the home. It will integrate video, high-definition TV, data, and voice, as shown in Figure 10-10. It is predicted that the bit rate will be between 150 and 600 Mbps.

10.10 SYNTRAN

SYNTRAN, an acronym for *synchronous transmission*, drops and adds DS-1 lines from a T3 line. The unique aspect of this standard is that the drop and add are accomplished without demultiplexing and remultiplexing all the channels. Each SYNTRAN terminal handles 28 DS-1 lines, with a data rate of 64 kBps.

10.11 FDDI AND FDDI-II

ANSI standard X3T9.3, the *fiber distributed data interface* (FDDI), is a combination of standards. FDDI combines the dual token-passing ring with IEEE standard 802.2, the logical link control (LLC) layer. The data rate is from FDDI, which is a 100 Mbps token-passing ring. The wavelength of operation is 1300 nm employing light emitting diodes. Multimode graded-index fiber is used. The total link length can be 200 km, with stations or nodes spaced 2 km apart. A maximum of 500 stations can be used.

FDDI consists of three layers: a physical layer, data link layer, and station management layer. These layers are controlled through software and hardware. The physical layer (PL) is separated into a physical medium dependent (PMD) and a physical layer (PHY) protocol. The PMD defines the transceivers, cables, connectors, and code requirements. The PHY is the link between the PMD and the data layer. The PHY includes such things as the clock synchronization, and encoding and decoding the frames. FDDI is shown in Figure 10-11. The data link layer (DLL) controls access to the cable

	8 bits		
Preamble	Starting delimiter	Frame control	End delimiter

Token

	8 bits		16 or 48 bits	16 or 48 bits		32 bits		
Preamble	Starting delimiter	Frame control	Destination address	Source address	Information	Frame check	End delimiter	Frame status

Frame

Figure 10-11 FDDI frame structure.

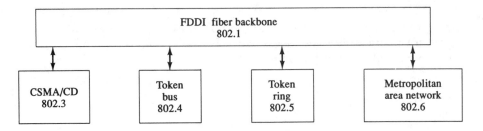

Figure 10-12 FDDI backbone.

and addressing information. This is like the IEEE 802.2 specification. The station management (SMT) layer is the manager of FDDI. The SMT monitors alarms, recovery, control, and management of the ring.

FDDI-II will carry voice and imaging. This standard will be able to carry data from color faxes, databases, x-ray, magnetic resonant imaging (MRI), voice, and video. FDDI-II uses the same fiber optic cable, path length, number of stations, and bit rate. The only addition to FDDI is the layering, which includes a hybrid ring control. FDDI and FDDI-II will become the backbone of wide-area networks, as shown in Figure 10-12.

10.12 SONET

SONET, an acronym for *synchronous optical network*, is the ANSI standard T1.105. At one time this was the only standard specifically designed for fiber optics. It provides for a standard synchronous interface between the optical and electrical interfaces. SONET defines the signal in a synchronous frame structure for multiplexing digital data. Procedures to link equipment from one manufacturer to any other, such as IBM to Digital Equipment Corporation computers, are defined.

The synchronous frame structure is called the synchronous transport signal, level 1 (STS-1). STS-1 has a bit rate of 51.84 Mbps with a frame structure of 90 columns by 9 rows by 8 bits. One entire row is transmitted every 125 μs. The first three columns in the frame are overhead bytes, while the remaining 87 bytes carry the synchronous payload envelope. The STS-1 can carry low-level signals such as DS-1, DS-1C, and DS-2. It can also carry DS-3 rate signals (44.736 Mbps). The frame structure is shown in Figure 10-13.

SONET is divided into layers called the physical layer, section layer, and line layer. The physical layer sets up the transport of bits as optical or electrical pulses. The section layer deals with the transport of the frame that uses the physical layer. The line layer provides for synchronization and multiplexing for the physical path layer.

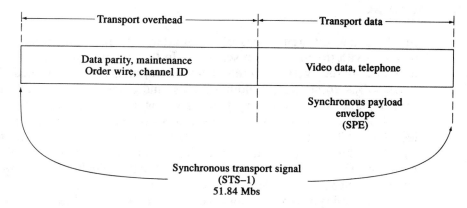

Figure 10-13   SONET frame structure.

10.13 BROADBAND SYSTEMS

Broadband technology is the hope for fiber use in the home. It will provide voice, video, and data on a narrow band while providing pay-for-view, one-way addressing, and two-way services (interactive classrooms). A possible scenario of broadband services is shown in Figure 10-14.

The cable TV (CATV) industry has already tried to employ broadband technology using fiber. In the HI-OVIS trial in Japan, discussed in Chapter 1, it was found that homeowners would pay only for basic services. Using fiber optics simply complicated the installation process. CATV has pushed to put fiber into new subdivisions before the houses are completed, to lower installation costs. The public will have to demand more of these types of services so that the prices of installation and terminal equipment will go down.

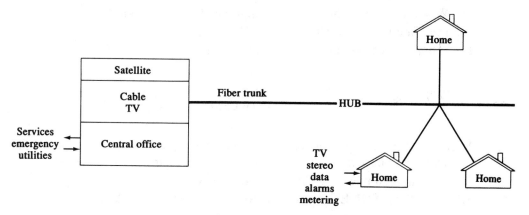

Figure 10-14   Broadband services.

10.14 SUMMARY

In Chapter 9, different types of transmitters and receivers were discussed, together with the modulation schemes. In this chapter, by using those transmitters and receivers, different network topologies are designed. Various ways to link computers, buildings, areas, and cities have emerged. The four primary types of network topologies are star, ring, bus, and tree.

Local area networks are used to link computers and buildings that are less than 1 kilometer away. Metropolitan area networks are used to transport signals within a city, campuses, buildings, and between LANs. Wide-area networks use long-haul trunking, typically for the telephone industry, to link cities together.

Standard types of fiber optic local area networks and wide-area networks are described, such as Ethernet, SONET, and FDDI. Each has its own merits and complications. Each is trying to emerge as the winner for fiber optic use in the home.

QUESTIONS

1. What was the only standard written specifically for fiber optics?
2. What is the primary reason to use ISDN?
3. In your opinion, which one of these networks seems more likely to succeed, and why?
4. What complications exist with broadband networks?
5. Which network does not use multiplexing to drop and add DS-1 signals?
6. What are the four major topology types?
7. What is the simplest network to have, and why is it not practical in a city environment?
8. Which network would be used that was less than 1 km?
9. Describe the token ring.
10. What advantage does FDDI-II have over FDDI?

PROBLEMS

1. If there is a star bus using a transmissive coupler, how much loss would there be if there were 10 stations? (Refer to Chapter 5.)
2. Draw an example of a physical ring/logical star.
3. How many SONET rows can be transferred in 1 minute? How many frames?

4. An FDDI network has been installed and there are some delays in transmitting the signal. There are 200 stations, with a 100 μs delay at each station. What is the total delay for all stations, and is it less than 500 ms?

5. Forty SYNTRAN terminals are set up to handle a maximum of 28 DS-1 channels. What is the total bit rate?

11

System Design

The student will be able to:

- Understand trade-offs in components and fiber.
- Calculate power budgets for a fiber optic link.
- Perform risetime analysis.
- Calculate repeater spacing.
- Calculate power budgets and number of subscribers for LANs.

11.1 TYPICAL FIBER OPTIC SYSTEM

Now that all the components of a fiber optic system have been discussed, the stage has been set to build an entire system. The system design boils down to choices of components and fiber. The overriding factors involved in deciding what system to use depend on the bandwidth of the modulation scheme and the distance of the fiber route. Elementary fiber optic design is essential to knowing if the system will work. The choice of the optical source (LED or laser), the optical fiber (multimode or single-mode), and the optical detector (PIN or APD) is discussed. Calculation of performance parameters is also covered, to provide more insight into complete system design.

Figure 11-1 shows a typical fiber optic communications system. The input signal is processed through some type of coding or modulation scheme.

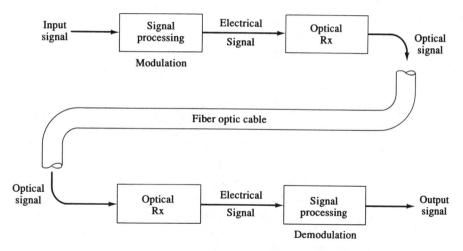

Figure 11-1 Typical fiber optic communication system.

The electrical signal is converted to light at the optical transmitter. As the light travels through the fiber, it must go through splices and connectors to arrive at the optical receiver. At this point the light is converted back into an electrical signal that is decoded or demodulated to form the output signal. The output signal should closely resemble the electrical signal that was originally the input to the system.

11.2 DESIGN METHODOLOGY

Fiber optic system design is not a trivial task. First, the user's needs must be considered. Factors such as cost, distance, bandwidth, number of channels, environment, and data rate must be defined before the design work can begin. Two or three designs may emerge as practical solutions to the same problem. These designs must be weighed against what the user is willing to pay and what is needed to do the job. Also, future use of the system must be considered.

11.2.1 System Considerations

Digital, analog, audio, or video? This is the first consideration. Once the type of signal has been established, the bandwidth must also be set. Other aspects, such as bit error rate or signal-to-noise ratio, distortion, and crosstalk, all aid in focusing on the type of transmitter or receiver that will be used.

11.2.2 Sources

The choice of the source type (LED or laser) is based primarily on distance and bandwidth. Laser sources have higher bandwidth and can be modulated

---

up to 6 GHz. For smaller bandwidth requirements, LEDs are quite suitable. Both LEDs and lasers are available for operating at 0.8, 1.3, and 1.5 μm wavelengths. Some benchmarks to look at are:

1. LEDs should be used for low to moderate bandwidth requirements (0 bps to 10 Mbps).
2. LEDs are not well suited for studio-quality analog video transmission.
3. LEDs have a longer lifetime.
4. Laser diodes are used for the higher digital data rates (>565 Mbps).
5. Laser diodes require extra circuitry to have constant power output over a given temperature range. This adds to the cost of the laser system.
6. Lasers must be biased at 20 to 100 mA to maintain the lasing action.
7. Lasers can couple more energy into a fiber than a LED can, due to their radiation pattern being much smaller than that of a LED. This means that the laser can transmit a greater distance and thus is suited for single-mode fiber.

LEDs cost from $0.25 to $100. The tighter the specifications, the costlier the source. ILDs are costlier, from $10 to $1000. Lasers have a beginning cost of $200 to $500, with upper costs of $10,000 or more. Nevertheless, the price has to be paid to obtain the power wanted for the link.

11.2.3 Fiber

Fiber considerations are usually driven by the length of the system. Following are the fiber types used for the link distances:

1. Multimode fiber is used for short distances, usually less than 2 km.
2. Single-mode is used for longer lengths.
3. Plastic is used for very short lengths such as 1 to 2 m.

The attenuation of the fiber ranges from 0.2 dB/km (single-mode) to 100 dB/km (plastic). Therefore, depending on the length of the link, the attenuation of the fiber becomes an important consideration.

Although attenuation can be a major loss, future growth of the cable plant is also important. The number of fibers in a cable can be as little as one, with no fiber count limit. The minimum number of fibers in a cable should be two, one for the signal and one for redundancy. The maximum number of fibers per cable is 12 fibers per buffer tube. Over 144 fibers can be put in one cable configuration.

The cost of graded-index multimode fiber is about $0.30 per meter, single-mode fiber is about $0.10 per meter, and plastic fiber is roughly $0.02 per meter. Be careful of prices quoted, because some manufacturers quote on uncabled fiber. If special cabling is required, the price of cabling becomes significant. If a standard cable configuration is used, which includes buffer tubes, central member, and so on, take the fiber price, multiply it by the number of fibers in the cable, and add about $1 per meter. That will provide an estimate of the total fiber cable cost.

11.2.4 Detectors

The choice of optical detector is based on responsivity, dark current, rise times, reverse voltages, and price. Some other practical choices to consider are as follows:

1. High bandwidth systems use the APD.
2. Low bandwidth systems use the PIN photodiode.
3. APDs must have some added circuitry to maintain the reverse voltage between 20 and 200 V over a certain temperature range.
4. APDs have noisier systems, so added circuitry is needed, at added cost.
5. An automatic gain control (AGC) circuit can be used to maintain a constant signal amplitude for the APD, but at a much higher cost in circuitry.

11.2.5 Connectors

Connectors are a function of the type of application and the fiber cable choice. The most popular types of connectors used for all applications are the SMA and the biconical. The biconical connector is quite large; hence if there is a space constraint, the SMA connector is the preferred choice. Connector costs vary from a few dollars to hundreds, depending on the precision machining needed to align and secure the fiber. SMAs are used for short distances and biconics for long-haul systems.

11.2.6 Installation Costs

Installation is probably the most costly part of the system, especially for a long haul system. This fact fits into the context of this chapter because it is not always performance trade-offs that must be met, but cost. Systems are designed to fit within a given budget. A typical system is defined in Figure 11-2. The costs associated with the equipment and installation of one such system are shown in Table 11-1. The table does not include the costs for connectors, jumpers, barrier signs, renting backhoes, or providing for an

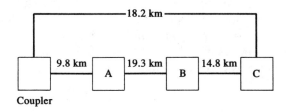

Coupler

Figure 11-2 Digital system design.

TABLE 11-1 Costs for Equipment and Installation

Item	Cost	Quantity	Extended Cost
45-Mbps Tx/Rx	$11,000	12	$132,000
Optical repeater	$13,000	1	13,000
Six-fiber cable	$2.80/meter	62 km	173,000
Rented fiber splicer	$400/week	1	400
Technician	$300/day	2 for 3 days	1,800
			$320,200

engineer to come to the installation. All these prices pertain to the company and installation type. Obviously, the cost of an installation can rise quickly. Sometimes a system can be redesigned, resulting in significant savings.

11.3 ANALYSIS

The pieces of the puzzle are beginning to fit together, bringing a system configuration to reality. There are also some rule-of-thumb systems that can be used to start with.

1. Very short lengths (less than 100 m)
   Use visible LEDs
       plastic fiber
       silicon detectors with low sensitivity
2. Short local area networks (1 to 2 km)
   Use short wavelength (850 nm) LEDs
       graded-index fiber
       InGaAs PIN receivers with medium sensitivity
3. Medium length systems (5 to 10 km)
   Use long wavelength (1300 nm) ILDs or LEDs
       graded-index or single-mode fiber
       InGaAsP PIN or APD detector receivers with medium sensitivity
4. Long haul systems (40 to 50 km)
   Use long wavelength (1550 nm) ILDs or laser diodes
       single-mode fiber
       InGaAsP APD detectors with high sensitivity

Two types of analysis are required for the design of a fiber optic system: power throughput analysis or loss budget and risetime analysis. Repeater spacing must also be calculated for long-haul links. Examples of these are given below. Designs are done using tables. To take full advantage of the design technique, an outline of the various calculations needed for the tables are given.

11.4 POWER THROUGHPUT ANALYSIS (LOSS BUDGET)

Table 11-2 is the worksheet used when analyzing power throughput. The following list explains the worksheet.

1. Determine the required bandwidth or bit rate.
2. Determine the total length of the system from transmitter to receiver.
3. If analog, determine the signal-to-noise ratio. The larger the SNR, the better.
4. If digital, determine the bit error rate. Typical BERs are 10^{-9} or 10^{-12}.
5. Determine the attenuation of all the fiber types used in the system.
6. Calculate the fiber bandwidth:

$$\text{fiber bandwidth} = \frac{\text{total fiber bandwidth}}{\text{link distance}} \qquad (11\text{-}1)$$

7. Write down the source type and average source output power, P_s.
8. Write down the detector type and receiver sensitivity, P_r.
9. Calculate the total power margin, which is equal to average output power minus the receiver sensitivity.

TABLE 11-2 Power Throughput Analysis

```
LOSS BUDGET
Required bandwidth or bit rate: _____
Required distance:              _____
Required SNR or BER             _____
Fiber type:                     _____
Total fiber bandwidth @ _____ MHz-km: _____ MHz
Source type: _____   Avg. source output power: P_s _____ dB
Detector type: _____   Receiver sensitivity: P_r _____ dB
                        TOTAL MARGIN (P_s − P_r) _____ dB
                        Source coupling loss: L_0 _____ dB
                        Total fiber loss @ dB/km _____ dB
Number of connectors: _____   Total connector loss
                @ _____   dB/connector _____ dB
Number of splices:    _____   Total splice loss
                @ _____   dB/splice      _____ dB
                        Detector coupling loss _____ dB
                        Temperature degradation _____ dB
                        Time degradation _____ dB
                        TOTAL ATTENUATION _____ dB
EXCESS POWER (TOTAL MARGIN −
                TOTAL ATTENUATION) _____ dB
```

10. The source coupling loss is 1 dB. This may seem large, but it gives the designer some margin for error.

11. Calculate the total fiber loss, which is equal to the fiber loss (dB/km) multiplied by the length. If different types of fiber are used (i.e., aerial and burial), simply add the multiplications together:

[fiber loss (aerial) × length] + [fiber loss (burial) × length]

12. Calculate the connector loss, which is equal to the number of connectors multiplied by the connector loss (dB). Connector loss is derived from the manufacturer's data sheets. Connector loss values are typically less than 1 dB.

13. Calculate the splice loss, which is equal to the number of splices multiplied by the splice loss (dB). Typical splice loss values are 0.01 to 0.02 dB.

14. Detector coupling loss is 1 dB. This gives the designer some margin for error.

15. Temperature degradation is 3 dB. This is a standard.

16. Aging or time degradation is 3 dB apiece. This is a standard.

17. Calculate the total attenuation by adding the source coupling loss, total fiber loss, connector loss, splice loss, detector coupling loss, temperature degradation, and time degradation to determine the total system attenuation.

18. Calculate the excess power, which is equal to the total margin minus total attenuation.

If the difference is negative, the receiver sensitivity should be changed to a larger power margin. If a larger power margin cannot be found, try changing the cable; for instance, use single-mode instead of multimode. If the fiber has been changed, fewer splices or connectors may be needed. If the excess power margin is equal to or greater than 4 dB, use the system.

Example 11-1

A medium range system is being designed. The system is a 10-km digital link with a bit rate of 45 Mbps and a BER of 10^{-9}. The source is an ILD with 3.25 mW of input power. The output power of the ILD is 5 dB. Multimode fiber with an attenuation of 3 dB/km will be used. Total fiber bandwidth is 300 MHz-km. A silicon APD with a sensitivity of -53 dB is the receiver. There are two connectors and three splices in the system.

LOSS BUDGET: Medium Range System
Required bandwidth or bit rate: 45 Mbps
Required distance: 10 km

```
Required SNR or BER 10⁻⁹
Fiber type: graded index 3 db/km
Total fiber bandwidth @ 300 MHz km: 30 MHz
Source type: ILD 3.25 mW    Avg. source output power: Pₛ    5      dB
Detector type: Si APD               Receiver sensitivity: Pᵣ  −53      dB
                                     TOTAL MARGIN (Pₛ − Pᵣ)   58      dB
                                     Source coupling loss: L₀    1      dB
                                     Total fiber loss @ 3 dB/km  30     dB
Number of connectors: 2              Total connector loss
                  @ 1                dB/connector                2      dB
Number of splices:    3              Total splice loss
                  @ 0.02             dB/splice                        0.06 dB
                                     Detector coupling loss       1      dB
                                     Temperature degradation      3      dB
                                     Time degradation             3      dB
                                     TOTAL ATTENUATION        40.06 dB
         EXCESS POWER (TOTAL MARGIN − TOTAL ATTENUATION) 17.94 dB
                                     (58 − 40.06)
```

The system designed in Example 11-1 will work with a little to spare.

11.5 RISETIME ANALYSIS

Risetime analysis permits calculation of the bandwidth for a fiber optic system. To calculate the total risetime, square the risetimes of the major components most affected by the delays: the source, multimode and material dispersion, and the detector. Sum the squares. Take the square root of the sum and multiply by 1.1. The multiplication factor is to add a 10% tolerance level.

$$\text{System risetime} = 1.1 \sqrt{T_1 + T_2 + T_3 + \cdots} \qquad (11\text{-}2)$$

The calculated risetime must always be less than the required system risetime. To determine if the system risetime is adequate for the signal, the modulation schemes must be considered.

1. For NRZ, the total required system risetime should be no more than 70 percent of the system risetime.
2. For analog, it should be no more than 35 percent of the risetime.
3. For PCM, the risetime should be calculated as follows:

$$\text{sampling rate} \times \frac{\text{pulse separation}}{\text{pulse width} \times \text{bandwidth}} \qquad (11\text{-}3)$$

Table 11-3 shows a worksheet to use for the risetime analysis.

TABLE 11-3 Risetime Analysis

REQUIRED SYSTEM RISETIME: Mbps   or   ns

Digital NRZ: 0.35/baud rate

PCM: $\left[\text{sampling rate} \times \dfrac{\text{pulse separation}}{\text{pulse width} \times \text{bandwidth}} \right]$

REQUIRED FIBER LENGTH, TYPE:

DESCRIPTION	RISETIME (ns)	RISETIME SQUARED
Source type:		
Total fiber risetime due to multimode dispersion @ ___ ns/km		
Total fiber risetime due to material dispersion @ ___ ns/km		
Detector type:		
Receiver (if analog):		

SUM OF SQUARES:

System risetime (square root of sum):   ns

Analog system (3 dB bandwidth) (0.35/rise time):   MHz

Example 11-2

Given the following parameters, Table 11-3 is set up to calculate the risetime.

$\lambda = 820$ nm

Fiber type: graded-index multimode

Baud rate: 45 Mbps

Detector risetime: 3.0 ns, from specifications

Transmitter risetime: 5.0 ns, from specifications

Modulation technique: NRZ

Link length: 10 km

Modal dispersion: 0.121 ns/km, from manufacturer

Material dispersion: 0.328 ns/km, from manufacturer

REQUIRED SYSTEM RISETIME: 45   Mbps   or 7.78 ns

REQUIRED FIBER LENGTH, TYPE: 10 km graded index

DESCRIPTION	RISETIME (ns)	RISETIME SQUARED
Source type: ILD	5	2.5×10^{-17}
Total fiber risetime due to multi-mode dispersion @ 0.121 ns/km	1.21	1.46×10^{-18}
Total fiber risetime due to material dispersion @ 0.328 ns/km	3.28	1.075×10^{-17}
Detector type: PIN	3	9×10^{-18}
Receiver (if analog):		

SUM OF SQUARES: 4.622×10^{-17}

System risetime (1.1 × square root of sum): 6.798 × 1.1 = 7.478 ns

The calculated system risetime should always be less than the required system risetime. If it is not, the best way to fix the problem is to get a faster source, different cable, or faster detector.

11.6 REPEATER SPACING ANALYSIS

Use of Table 11-4 is only for long haul links. The analysis is as follows:

1. Calculate the power (dB) into the fiber. See Appendix A for conversion of watts to dB.
2. Calculate the receiver sensitivity or derive it from manufacturer's specifications.
3. Calculate the loss due to the connectors, which is equal to the number of connectors multiplied by 1 dB/connector.
4. Choose a system margin, or use the one calculated from the power throughput analysis. If there is lack of information, use at least 5 dB.

5. Calculate the attenuation between repeaters by adding the steps 1 to 4 on page 185.
6. Find the fiber attenuation from the specification sheets.
7. Calculate the splice attenuation by multiplying the number of splices by the loss per splice: for example, 3 splices/km × 0.02 dB/splice.
8. Reserve repair splices are 0.1 dB/splice.
9. Calculate the cable attenuation by summing the fiber attenuation, splice attenuation, and reserve repair splices.
10. Calculate the maximum repeater spacing by taking the attenuation between repeaters divided by the cable attenuation.

TABLE 11-4   Repeater Spacing Analysis

REPEATER SPACING ANALYSIS	
Power into the fiber (mW)	_____ dB
Receiver sensitivity Mbps	_____ dB
Attenuation due to connectors (dB/termination)	_____ dB
System margin	_____ dB
ATTENUATION BETWEEN REPEATERS	_____ dB
Fiber attenuation @ ___ nm	_____ dB/km
Splice attenuation (___ splice/km)	_____ dB/km
Reserve repair splices	_____ dB/km
CABLE ATTENUATION	_____ dB/km
Maximum repeater spacing (attenuation between repeaters/cable attenuation)	_____ km

Example 11-3

The following worksheet shows an example of calculating repeater spacing for a laser diode source, multimode graded-index fiber, and an avalanche photodiode as receiver.

REPEATER SPACING		
Power into the fiber (2 mW)	+3	dB
Receiver sensitivity 100 Mbps	−50	dB
Attenuation due to connectors (1 dB/termination)	4	dB
System margin	5	dB

System Design   *Chap. 11*

ATTENUATION BETWEEN REPEATERS	−38	dB
Fiber attenuation @ 850 nm	3.0	dB/km
Splice attenuation (0.3 splice/km)	0.3	dB/km
Reserve repair splices	0.3	dB/km
CABLE ATTENUATION	3.6	dB/km
Maximum repeater spacing (attenuation between repeaters/cable attenuation)	10.55 km	

The calculated repeater spacing of 10.55 km is about right for this type of system.

11.7 BUDGETS FOR LANS

Each basic topology for a LAN has to be calculated separately. A passive star configuration will differ from an active ring or active star configuration. The round-trip time delay must also be calculated.

11.7.1 Passive Star Topology

A passive star topology is depicted in Figure 11-3. Each station's transmitter couples power into the star mixer. Each station's receiver accesses the star mixer. The star mixer receives all transmitted signals and distributes the signal equally among all receiver ports. The star output is 1 dB/amount of station's power less than the original signal. The power budget is based on the following equation:

$$P_s - P_{r\min} - M > (L_{ct} - L_{cr} - 2L_c) - L_{star} + D(L_f) \qquad (11\text{-}4)$$

P_s is the transmitted power before the connector, in dBm

$P_{r\min}$ is the minimum receiver power, in dBm

M is the operating margin or excess power, in dB

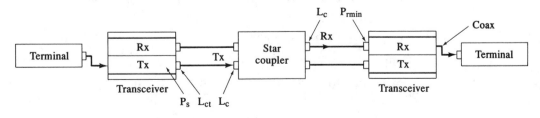

Figure 11-3 Passive star topology.

L_{ct} is the connector loss at the transmitter, in dB

L_{cr} is the connector loss at the receiver, in dB

L_c is the connector loss at the coupler, in dB

L_{star} is the splitting loss of the coupler: 10 log(1/no. ports)

D longest distance between stations, in km

L_f is the fiber loss, in dB/km

Equation (11-4) states that the system gains must be greater than the system losses. The following equation is used to calculate the number of stations that the coupler can have:

$$L_{IL} = L_{coupler} + 10(N) \qquad (11\text{-}5)$$

where

$$N = \frac{P_t - P_r - M - L_{ct} - L_{cr} - 2K_c - L_{coupler} - D(L_f)}{10} \qquad (11\text{-}6)$$

$$\text{number of stations} = 10^N \qquad (11\text{-}7)$$

L_{IL} is the total insertion loss of the coupler.

All other nomenclature is as defined earlier.

Example 11-4

Given the following parameters, the number of stations can be calculated.

  λ = 820 nm

  Fiber type: multimode graded index

  Bandwidth: 10 Mbps

  LED transmitter power: 0 dBm

  Receiver power: -30 dBm

  Link length: 500 m

  Margin: 3.0 dB

  Connector losses, transmitter and coupler: 1 dB each

  Connector losses, receiver: 0.5 dB

  Fiber loss: 3 dB/km at 820 nm

  Coupler loss: 5 dB

$$N = \frac{0 - (-30) - 3 - 1 - 2(1) - 0.5 - 5 - 0.5(3)}{10} = 1.7$$

$$N = 10^{1.7} = 50.118$$

This means that the star coupler can handle a maximum of 50 stations. Since most large couplers can only be built with 64 stations, it is safe to assume that this LAN will work.

11.7.2 Active Ring and Star Topology

These types of topologies are point to point between stations. Figure 11-4 shows both topologies. In the ring topology, each station is connected from transmitter to receiver by a duplex cable. A token passing protocol is used to share the data. For the star topology, each station is connected to an $N \times N$-port active star. It is point to point between the station and the active star. A CSMA/CD protocol is used.

The following equation is used to calculate the power budget:

$$P_t - P_{r\min} - M > L_{ct} - L_{cr} + D(L_f) \tag{11-8}$$

P_t is the transmitted power before the connector, in dBm

$P_{r\min}$ is the minimum receiver power, in dBm

M is the operating margin or power penalties, in dB

L_{ct} is the connector loss at the transmitter, in dB

L_{cr} is the connector loss at the receiver, in dB

D is the longest distance between stations, in km

L_f is the fiber loss, in dB/km

The power budget again states that the system gains must be greater than the system losses. The losses can be improved to get a lot of ports; the drawback is that the practical coupler can handle only 64 ports.

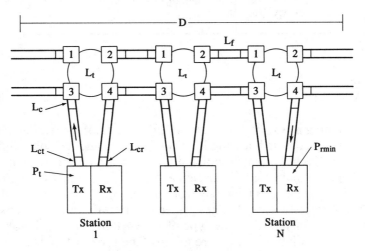

Figure 11-4 Active ring and star topology.

11.7.3 Round-Trip Delay

The round-trip delay calculation is used only for collision detection. Recall from Chapter 10 that any station transmitting a packet must be able to sense a collision with any other station on the network. To calculate the round-trip delay, the total round-trip delay must be smaller than the packet length.

Example 11-5

Given the following parameters, the round-trip delay will be calculated.

Transmission path delay (μs):

Electronic circuits	2.15
Cable propagation	
Fiber	21.88
Coax to terminal	0.52
	24.55

Return delay (μs):

Electronic circuits	2.25
Cable propagation	
Fiber	21.88
Coax to terminal	0.52
	24.65

Transmission path delay (μs): 49.20

If using a minimum SONET packet length of 51 μs, all the collisions will be sensed by the stations.

11.8 SUMMARY

Designing a system takes a lot of knowledge of the available components and their price. Just knowing how the link is going to look will help in the decisions. Even a point-to-point link is not a trivial case if it is in a network. There are many ways to design a system, including the rule-of-thumb method and/or listening to what has worked before. At least this is the starting point, but ultimately a loss budget, risetime analysis, and repeater spacing will have to be calculated. If some type of network topology is involved, a whole new set of calculations will be required. The type of topology and the number of stations that can be put on the network are utilized.

11.9 EQUATION SUMMARY

Fiber bandwidth:

$$\text{fiber bandwidth} = \frac{\text{total fiber bandwidth}}{\text{link distance}} \qquad (11\text{-}1)$$

System risetime:

$$\text{system risetime} = 1.1 \sqrt{T_1 + T_2 + T_3 + \cdots} \qquad (11\text{-}2)$$

PCM risetime:

$$\text{sampling rate} \times \frac{\text{pulse separation}}{\text{pulse width} \times \text{bandwidth}} \qquad (11\text{-}3)$$

Passive star power budget:

$$P_s - P_{r\min} - M > (L_{ct} - L_{cr} - 2L_c) - L_{star} + D(L_f) \qquad (11\text{-}4)$$

Number of stations for passive star coupler:

$$L_{IL} = L_{coupler} + 10(N) \qquad (11\text{-}5)$$

where

$$N = \frac{P_t - P_r - M - L_{ct} - L_{cr} - 2L_c - L_{coupler} - D(L_f)}{10} \qquad (11\text{-}6)$$

$$\text{number of stations} = 10^N \qquad (11\text{-}7)$$

Power budget for active ring and star topology:

$$P_t - P_{r\min} - M > L_{ct} - L_{cr} + D(L_f) \qquad (11\text{-}8)$$

QUESTIONS

1. What type of source should be used for a short distance link?
2. Which source, laser, or LED can couple more energy into the fiber?
3. Which type of system, low or high bandwidth, uses APDs?
4. Which detector is noisier, the APD or the PIN?
5. Why is the round-trip delay calculated for collision detection?

PROBLEMS

In Problems 1 to 4, calculate the power throughput analysis for each digital system.

1. Very short range system
30 m link at 10 MHz with plastic fiber

Source type: LED with power out of -10 dB
Detector type: PIN with receiver sensitivity of -39 dB
Source coupler loss: 5 dB
Plastic fiber attenuation: 100 dB/km
No splices
Two connectors at 3 dB/connector
Aging and temperature degradation: 5 dB
Detector coupler loss: 5 dB
BER: 10^{-9}

2. Short range system
   1-km link at 100 MHz with plastic-clad silica fiber
   Source type: LED with power out of 5 dB
   Detector type: PIN with receiver sensitivity of -40 dB
   Source coupler loss: 5 dB
   Plastic-clad silica fiber attenuation: 5 dB/km
   One splice at 1 dB/splice
   Four connectors at 1 dB/connector
   Detector coupler loss: 5 dB
   BER: 10^{-9}

3. Medium range
   10 km link at 100 MHz with graded index fiber
   Source type: LED with power out of 0.0 dB
   Detector type: PIN with receiver sensitivity of -36.5 dB
   Source coupler loss: 5 dB
   Graded-index fiber attenuation: 1.0 dB/km
   Four splices at 0.3 dB/splice
   Six connectors at 0.75 dB/connector
   Aging and temperature degradation: 6 dB
   Detector coupler loss: 5 dB
   BER: 10^{-9}

4. Long range system
   40 km link at 420 MHz with single-mode fiber
   Source type: ILD with power out of 10 dB
   Detector type: APD with receiver sensitivity of -28.8 dB
   Source coupler loss: 7.5 dB
   Single-mode fiber attenuation: 0.3 dB/km
   Ten splices at 0.1 dB/splice
   Four connectors at 0.2 dB/connector
   Aging and temperature degradation: 6 dB
   Detector coupler loss: 7.5 dB
   BER: 10^{-12}

5. Complete the risetime analysis using the data in Example 11-2, but use a 3-ns risetime for the LED. Will this system work?

6. Calculate the repeater spacing for the long range system.

7. Design a very short range application so that it will work.

8. Given a star topology and the following parameters, calculate the number of stations that the star coupler could have.

Transmitter power: 5 dB
Minimum receiver input power: −50 dB
Excess power margin: 6 dB
Loss of all couplers, two at star: 4 dB
Fiber loss for longest transmitter and receiver spacing: 25 dB
Splitting loss of coupler: 16 dB
Excess loss of coupler: 4 dB

12

Installation and Testing
of Fiber Systems

The student will be able to:

- Describe some installation requirements.
- Be familiar with enclosures and protection for the cable installation.
- Be acquainted with the test equipment and how it is used on a fiber optic system.

12.1 INSTALLATION

Since the fiber optic cable is designed to endure a certain level of mechanical structural strain, most of the conventional methods of electrical cable installation can be used, with a few modifications. These methods include pulling the cable into a duct, conduit, or trench; plowing the cable into the ground; and aerial, indoor, and water installations. The cable should not be placed in air-conditioning or ventilation ducts. If there is a fire in the duct, the polyvinyl chloride jacketing could burn and produce toxic gases.

Some questions that have to be answered before the cable can be installed are:

What is the outside diameter of the cable and the connectors?

How long is the cable run?

Are there access points, and what are the distances between those points?

Where are the bends located?

What is the conduit diameter, and how full is it?

Can water get in the conduit?

Is the conduit or trench below the frost line?

12.1.1 Pulling Cable in Ducts, Conduits, or Trenches

The cable can be pulled into a duct, conduit, or trench by hand. If a power winch is used, roller guides and sheaves must be used to protect the fiber. The two major rules of fiber optic cable installation are: **Do not pull directly on the fiber** and **Do not allow tight loops, kinks, or tight bends to occur.** There are some differences between pulling electrical cables and pulling fiber optic cables. Fiber optic cables may have connectors already installed at the ends, the pulling forces are smaller, and there is a minimum bend radius that must be adhered to.

A hand-formed pulling eye can be used if the cable count is small: less than two or three fibers per cable. On higher-count fibers, most cable installers have a special pulling tool that can be used. This pulling tool, called a "Chinese finger trap" or *Kellem's grip*, is shown in Figure 12-1. The pulling tape is attached to the strength member (usually the Kevlar). The connectors are wrapped in a thin layer of foam rubber and inserted inside the cable grip. The fiber cable and the grip are wrapped together with electrical tape. The cable is pulled using the grip rather than pulling on the connectors.

No matter how the cable is pulled, the tension must always be monitored. A tensiometer, or a dynamometer and pulley arrangement, is used for most hand pulling. If a power winch is used, a power capstan with an adjustable slip clutch is added. The slip clutch is set for the maximum load that the fiber cable can withstand during installation. If the tension is exceeded, the slip clutch will disengage, rendering the power winch unusable.

The cable should be lubricated continuously as it enters ducts or conduits. As the cable exits the conduit, it is coiled into a figure-eight pattern with 1 foot minimum loop. The figure-eight pattern prevents tangling and kinking of the cable. The cable is turned over and can now be pulled in the same manner. Most cable pulls start at or near the center point of the total run

Figure 12-1 Kellem's grip. (Courtesy of Kellems, a division of Hubbell.)

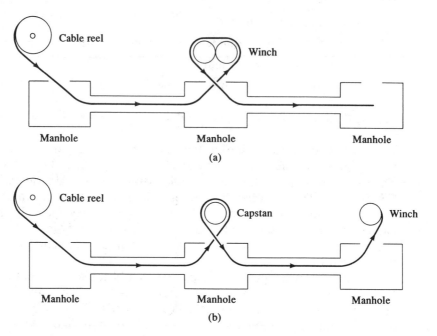

Figure 12-2 Pulling methods: (a) manual; (b) with winch.

or the center point between locales, as shown in Figure 12-2. This minimizes the amount of cable to be pulled.

If there are bends or angles in the cable route, such as at manholes or pullboxes, a pulley or wheel should be used. The pulley or wheel has a minimum diameter of 12 inches if under tension, and 4 inches if not. They are used so that the cable will not scrape against the conduit or exceed its bend radius.

12.1.2 Direct Burial Installation

The cable must be protected against frost, water intrusion, gnawing rodents, and earth movement. Cables can be buried directly in the ground by either plowing or trenching. The cable can be plowed in using the same methods as those used for electrical wire. Plowing involves a single operation whereby a trench is dug, the cable is laid, and the trench covered over. A paying-out system for the fiber optic cable is used. Again, the tension has to be monitored carefully so as not to exceed the design limits of the cable strength tension. The cable can be buried 36 to 48 inches deep, usually below the frost line, to help prevent environmental hazards. Loose tube cable is best suited for this application.

The fiber cable can also be drawn through a plastic conduit before the plowing operation starts. In this way, most of the stress is taken by the conduit. The only real drawback of this method is the high cost.

12.1.3 Aerial Installation

Most aerial installations of fiber optic cable are rigged along existing power poles or telephone lines. The cable must be supported on a messenger wire to prevent excessive strain. The aerial portion of the cable installation must be designed carefully, making sure of the route length, sag, pole spacing, and the total number of poles. The temperature range, ice loading conditions, and wind conditions must be known so that the fiber optic cable can be secured properly to the messenger wire. There is a special machine run along the messenger wire that lashes the fiber optic cable to the messenger wire. The installation is shown in Figure 12-3. The fiber optic cable should have a steel or fiberglass/epoxy rod as a stabilizing member.

12.1.4 Indoor Installation

Installing fiber optic cable in an indoor environment is similar to installing it outdoors in a conduit. An additional consideration is the environment in which the cable is installed: electrical equipment room, chemical or fuel storage areas, or subject to nuclear radiation. Flame-retardant cables should be used for interior installations. The cables are laid in conduits or trays in the plenum. The plenum is the space between the walls, ceiling, or raised floors. Fiber optic plenum cable has to meet the National Electrical Code® Article 770-7 for installation in air plenums without conduit.

12.1.5 Water Installation

For water installations, the considerations that must be taken into account are the tides and currents, water depth, and terrain on the bottom. Putting fiber optic cable in rivers, lakes, and streams requires more careful installation than with the other methods. A specific example of joining two cities for the phone company via underwater fiber optic cable occurred in 1982, when Seattle and Bellevue, Washington, were linked across a lake bed. Special considerations

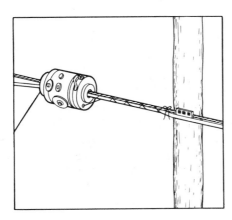

Figure 12-3 Aerial installation. (Courtesy of Cooper Industries, Inc., Belden Division.)

had to be used to run fiber optic cable in 60 m deep Lake Washington. Cost studies of copper versus fiber were evaluated, but because of the high-capacity bandwidth needed, the fiber cost per circuit mile was less. The cable chosen for this application contained 35 fibers with an underwater construction. The two 3.4 km cables would be laid 150 meters apart and joined at the shores to a 70 fiber duct cable.

To start with, divers secured the cable end into an underwater duct system. The rest of the cable was placed on a reel on a barge and was payed out as the barge moved slowly across the water. The speed of the barge was limited to 1 or 2 knots to reduce the cable strain. The fiber optic cable was payed out at a 45° angle to the surface of the lake, to provide a bend-free transition to the lake floor. When the barge reached the other end of the lake, the fiber optic cable was secured through the onshore duct system.

The cable for the Lake Washington application was a specially designed loose tube cable. Each of the fiber loose tubes contained seven fibers that were tightly buffered with an acrylate coating. The loose tubes were surrounded by a jacket that was filled with a moisture-resistant compound. The jacket was surrounded by steel tape and the whole cable was again surrounded by two layers of steel wire armoring.

Ocean installations are complicated and expensive. AT&T is updating the trans-Atlantic telephone link that will carry 80,000 telephone lines. The cable is designed to have six fibers, encircled with two layers of steel tape. The steel tape adds extra protection against shark attack. Shark attacks can cost the company billions of dollars. After one shark-feeding frenzy broke all the fibers, the telephone calls had to uplink to a satellite until it was repaired. This was an expensive week for transatlantic phone calls. To make repairs easier, AT&T has built a specially designed ship that will bury the fiber cables deeper at the seabed to prevent damage. The hold of the ship has three 25 ft steel cones from which the fiber will be payed out. A small laboratory is included for splicing the fiber. Buoys twice the size of a person are used for marking cable routes. Grappling hooks the size of refrigerators will be used to dredge up damaged cables. The ship also has a remote-controlled submarine, used to inspect cables after they have been laid.

12.2 ELECTRICAL BONDING AND GROUNDING

Twisted pair or coax have metallic elements and therefore must be bonded and grounded. For example, the telephone wires are bonded to power line grounds so that workers who might come in contact with the differences in potential would be protected. Grounding consists of deviating high voltage or current to a low resistance path to ground by a ground rod. When fiber optic cables include some metallic element, such as a steel central member, the metallic element must be bonded and grounded. If the cable is all dielectric, it is not necessary.

TABLE 12-1 Typical Optical Cable Specifications

Type of Installation	General Cable Characteristics	Tensile Strength for Installation (newtons)	Comments
Duct	Single jacket Filled loose tube	1000	Capable of being pulled in 1-km lengths; filled buffer prevents ice formation
Direct buried	Filled loose tube Stainless steel tape Rodent protection Rugged jacket Stranded fibers	1500	
Aerial	Stranded fibers Filled loose tube UV protected jacket	1000	For messenger wire installation; suitable for $\frac{1}{2}$-in.-thick ice load; designed for wide temperature range
Short-distance or computer link	Tight buffered Single jacket	600	Strength, flexibility, and small size

12.3 CABLE INSTALLATION AND CABLE TYPE

Table 12-1 summarizes the general characteristics and cable specifications for various applications. This table is based on general characteristics and typical mechanical specifications for SIECOR optical cable design.

12.4 HARDWARE

A splice closure (Figure 12-4) is used to protect cable splices. The fiber optic industry took a standard telephone splice closure and modified the internal part (organizer panel) to house the splices. An organizer panel (Figure 12-5) is used to neatly arrange the fibers and the splices in a stacked manner within the splice closure. The closures can hold from 12 to 144 splices.

Patch panels, shown in Figure 12-6, are used to reroute signals and fiber optic connections. One side of the panel, usually the back, has the fibers fixed. On the other side, the fibers can be connected and disconnected when needed. This patch panel is convenient in a large office building when tenants are transient. If all the tenants are connected to a smart environmental sys-

Figure 12-4 Splice closure and organizer. (Courtesy of AMP, Inc.)

Figure 12-5 Splice organizer. (Courtesy of Siecor Corporation, Hickory, North Carolina.)

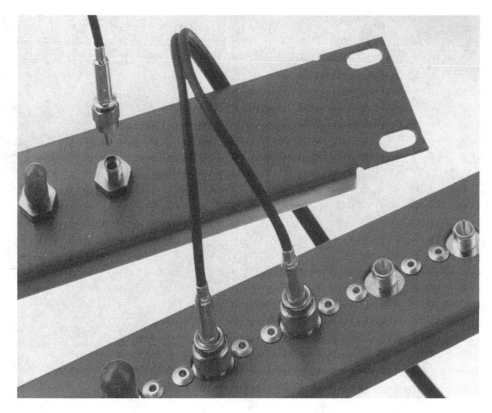

Figure 12-6 Patch panel. (Courtesy of AMP, Inc.)

tem or the mainframe computer, the ability to connect and disconnect, using the patch panel, saves on costs.

Junction boxes are similar to electrical outlets. These boxes are the link between the building wiring and the equipment. A short jumper cable runs from the junction box to the equipment (such as a terminal). The junction box shown in Figure 12-7 has a slatted front cover to prevent dust from entering the fiber optic connection. Figure 12-8 shows how all the hardware, from the outside of the building to the terminal on a worker's desk, can be implemented.

12.5 TESTING

After the system has finally been designed and built, it must be tested. The National Bureau of Standards and the Electronic Industry Association (EIA) have recommended many test procedures for use by the fiber optic community. RS-455 summarizes the importance of these tests and is a compilation of all the test procedures. Since most of the fiber cable measurements are completed at the manufacturer's site, these sections will focus on field testing for

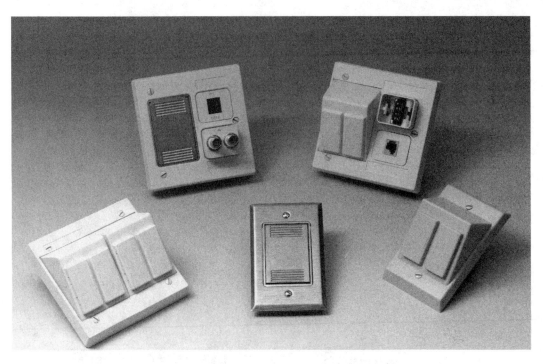

Figure 12-7 Junction box. (Courtesy of AMP, Inc.)

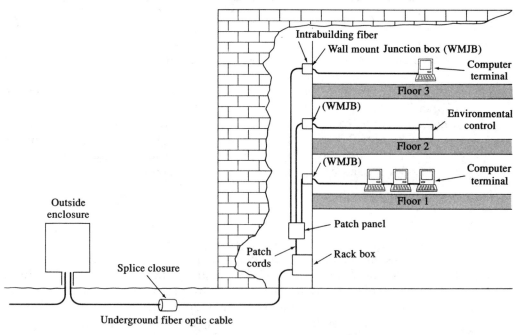

Figure 12-8 Hardware application. (Courtesy of Delmar Publishers, Inc., Albany, NY.)

the fiber optic system. Various instruments and testing methods are discussed.

12.6 OSCILLOSCOPES

The most common tool found in any lab is the oscilloscope, which can measure eye patterns and waveforms for the optical system. The oscilloscope must have a high bandwidth (40 MHz or more) to be usable for fiber optic transmissions. A reliable digital storage scope will always be useful to the fiber optic technician.

12.6.1 Eye Patterns

An eye pattern is shown in Figure 12-9. This is used to assess the quality of a digital link. To produce an eye pattern, a transmitter is connected to a pseudorandom data generator. The output is connected to the trigger of the scope. The display on the oscilloscope looks like an eye. The more open the eye, the better the transmission quality of the digital signal. If the eye is more closed, the more likely there are that errors are being transmitted. Several benchmarks of the eye pattern are used to measure the following:

1. Noise margin in the receiver: height of the central eye opening.
2. Jitter (variation in pulse timing): width of the signal band at the corner of the middle of the eye.

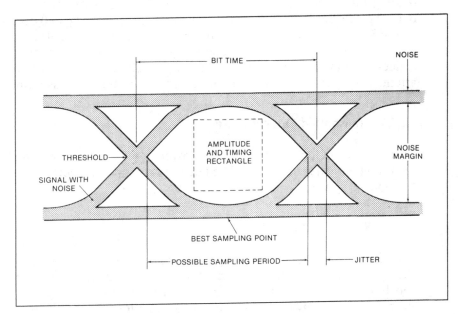

Figure 12-9 Eye pattern. (Courtesy of Sams: A Division of Macmillan Computer Publishing.)

3. Distortion in the receiver output: thickness of the signal line at the top and bottom of the eye.

4. Rise and fall time: transitions between the top and bottom of the eye pattern.

12.6.2 Cutback Method

The cutback method is used to measure fiber attenuation without connectors attached. A standardized light signal is launched into a long piece of fiber—on the order of 5 km. The resulting power out is read on the power meter. Then the same type of measurement is taken with a 0.5 m length of fiber. This reading represents launch power. Taking the ratio of the two power readings will provide a measure of decibel loss for the fiber. Dividing the decibel loss by the fiber length will give the dB/km specification for the fiber. This should match the manufacturer's specification sheet.

12.7 POWER METERS

These meters are used to measure power out of the system. The typical power meter shown in Figure 12-10 uses a silicon detector for the range 800 to 900 nm and a germanium detector for the spectral range 1300 to 1600 nm. The power meter detects and converts the light power to a dB, dBm, or μW reading on the meter. The most common reading is in dBm, which is the optical power referenced to 1 mW on a log scale. A reading of 0 dBm is 1 mW; -10 dBm is 100 μW. By using connector adapters and light sources of the same wavelength as the meter, the power loss measurements for connectors, splices, and the entire system can be made. The meter can be used for fiber attenuation measurements and connector insertion loss measurements.

12.7.1 Transmitted Optical Power

To measure the optical transmitter power, connect the power meter to the operating optical transmitter. The reading on the power meter is the transmitted optical power.

12.7.2 Received Optical Power

To measure the received optical power, connect the power meter to the end of the optical fiber, which is connected to the receiver. If the received power is too low or too high for the receiver sensitivity, the system will not work properly. If the power is too high, an optical attenuator can be inserted in the front of the receiver to limit the power. If the power is too low, a more

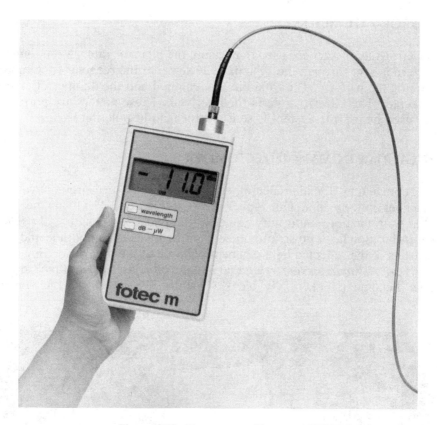

Figure 12-10   Power meter. (Courtesy of FOTEC, Inc.)

sensitive receiver, shorter cables, or higher transmitting power will be required.

12.7.3 Cable Attenuation

This test is used to measure the capability of the system to deliver the required power to the receiver. This test requires a stable optical source and an optical power meter. Take a short length of the fiber, and using the specified connector, couple the fiber to the optical source equipment. The other end of this short piece of fiber is connected to an optical power meter. The power is read directly from the power meter. Next, take the reading for the installed cable. The difference in the two readings should indicate the power lost due to the additional connectors, splices, repeaters, and fiber lengths. This difference should agree with the power throughput analysis within 1.0 dB. A larger discrepancy may show that the connections are inferior. Repolishing or replacing the connectors might solve the problem.

12.8 BIT-ERROR-RATE METERS

Bit-error-rate meters are used to measure the bit error rate. A randomized bit pattern is sent through the system. The signal at the receiver is compared to the original pattern. The total bits are counted and the number of errors are detected. The bit error rate is then derived. These meters are portable and are used before the system is sent out for field installation.

12.9 OPTICAL TIME DOMAIN REFLECTOMETER

The optical time domain reflectometer (OTDR) is the measurement workhorse of a fiber optic system. One type is shown in Figure 12-11. It can measure the fiber's attenuation, uniformity, splice loss, breaks, and lengths. By sending a short-duration laser pulse into one end of the fiber, the amplitude and characteristics of the reflected light can be monitored at the other end. Small reflections called *Rayleigh scattering* that occur as the laser pulse travels down the fiber become exponentially weaker as the distance from the source in-

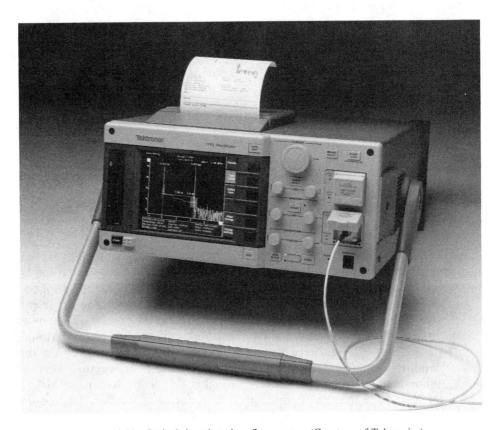

Figure 12-11 Optical time-domain reflectometer. (Courtesy of Tektronix.)

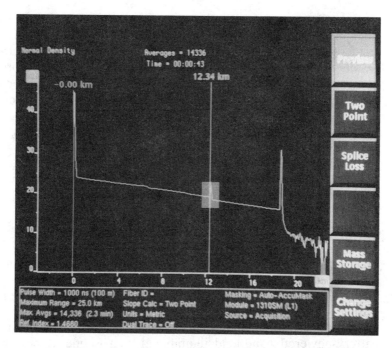

Figure 12-12 OTDR screen. (Courtesy of Tektronix.)

creases. An abrupt drop in the Rayleigh scattering indicates that there is a splice, connector, or fiber imperfection.

Most OTDRs have an oscilloscope type of CRT display. The difference between an OTDR and an oscilloscope is that the OTDR measures distance versus optical loss in dB. In Figure 12-12 the screen of an OTDR is shown. The amount of attenuation and the distance from the source can be clearly read off the OTDR screen. The optical decibel loss, the y axis, is automatically converted to a logarithmic scale that can display to the nearest 0.1 dB. The fiber distance, the x axis, can be measured to the nearest centimeter. The OTDR measures the distance along the fiber, not the length of the cable. For instance, if stranded cable is used, the actual fiber length is longer than the cable length. The slope of the curve indicates the fiber attenuation loss. The discontinuities on the curve represent connectors or splices. The drop at the discontinuity represents the amount of loss. The loss is only an average. If the loss is measured from both ends of the fiber, different readings will occur. To resolve the discrepancy, the average is taken for both readings.

In Figure 12-12, the distance of the total cable plant is 18.47 km. The fiber imperfection is less than 1.0 dB and occurs at 7 km from the beginning. The repeater occurs at 12.34 km out. The total cable attenuation or the slope of the line is 7.36 dB.

The OTDR has the advantage of taking measurements at only one end of the cable plant. The disadvantage is that for long distances, greater than 40 kilometers, the fibers must be measured from both ends.

Sec. 12.9   *Optical Time Domain Reflectometer*

A picture or trace produced from the OTDR screen is a record of each fiber in the cable plant. At the time of installation and splicing, an OTDR trace of each fiber is taken. The trace should be taken in both directions, so details such as microbends can be uncovered. Trace records are used to compare the existing cable plant with the originally installed cable. Degradation due to aging can be spotted and the cable can be replaced in that area. Copies of the traces, splice information, and the general report are kept at the installer's location for further reference. The customer can also request a copy of the report.

A word of caution: OTDR measurements are not exact. Any loss is an average. The operator must not rely absolutely on the readings, but rather, if a marginal loss, be able to redo the splice or connector.

12.10 MAGNIFYING GLASSES AND MICROSCOPES

Magnifying glasses and microscopes are used to detect scratches or defects on the ends of the optical fibers. If polishing of the fiber end during connectorization is not done to a mirror-like finish, a greater-than-average loss may occur through the connector. It is best to check the fiber end with at least a magnifying glass to ensure a good-quality connector joint. Some manufacturers sell a battery-powered hand-held illuminated microscope of power 30× to 100× which helps with satisfactory field inspection of the fiber end. An example of such a microscope is shown in Figure 12-13.

12.11 FLASHLIGHT TEST

The flashlight test is used to check for continuity of the fiber. The test is good for a short- to medium-length link test—about 1 mile maximum fiber span. A flashlight is used to check for continuity in the fiber. Take an ordinary flashlight and shine it into a cleaved or connectorized fiber. If you see light at the other end, the fiber is probably not a problem for the link. If outdoors, sunlight can be used to check for the continuity of the fiber. If the length is longer than a mile, the light at the other end may appear red in color. This is because of the filtering of the light within the fiber and does not indicate a fiber problem.

Warning: Although this test sounds easy to do, it is not always easy to see light from the fiber. Do **not**, out of desperation, hook one end up to the optical transmitter and look into the opposite end. This can cause permanent eye damage.

12.12 NETWORK TESTING

Many types of networks were described in Chapter 10, and all have to be tested in some manner. Testing the network involves testing the individual cable assemblies, the installed cable plant, the node electronics, and the entire

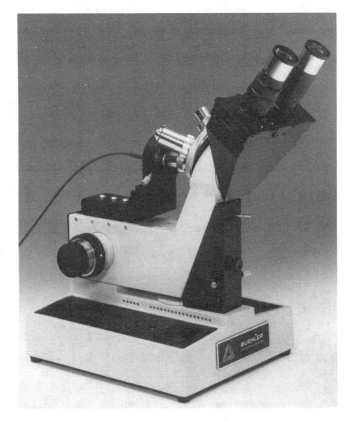

Figure 12-13 Fiber microscope. (Courtesy of Buehler Limited, Lake Bluff, Illinois USA.)

network system margin. This is a major task and must be completed in stages.

The first stage is to test each fiber individually for attenuation before it is installed. Typically, the installer can use the attenuation measurements after the cable has been installed. Second, the cable is tested for end-to-end attenuation. The method used is described in Section 12.7.3. The next step is to test the node electronics. The transmitters must have enough coupled optical power, and the receivers must have adequate sensitivity. To test the transmitter, the output power must be examined. A fiber optic power meter can be connected to the source and the power output read directly.

An optical receiver must be tested for optical signal input level and the quality (BER) of the signal using a *loopback test*. At the receiver end is an optical transmitter (for two-way transmission). By connecting an optical attenuator (lowers the level of the light out) to the transmitter and looping it back to the receiver, the BER can be measured. The transmitter is modulated with some digital signal. The receiver's optical input level is manipulated. The output of the receiver is connected to a BER tester. This loopback

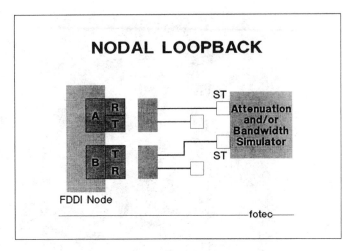

Figure 12-14 Loopback test. (Courtesy of FOTEC, Inc.)

method is depicted in Figure 12-14. This test takes, on average, 15 minutes to complete. Each individual channel must be tested.

The system margin test consists of applying a digital signal to the transmitter, attenuating the optical signal, and making BER test measurements at the receiver. A variable optical attenuator is used to attenuate the signal until the BER meets the minimum system performance test. When the degradation has been reached, the receiver's optical input signal level should be recorded. The attenuator is then removed and the received optical input is compared to the attenuated one. The technician working on the system must be able to make a judgment as to whether or not the signal is within the margins specified.

12.13 MAINTENANCE PRACTICES

So far, there is little or no maintenance on the fiber optic system. When the cable plant is cut, alarms sound on the electronic equipment, which tells the craftperson that there is no received power. The craftperson gets in a truck, preferably a four-wheel drive, and drives the route looking for someone digging. When the culprit is found, fines are levied on that company at the rate of $10,000 per minute. The time is counted from when it was cut to when the signal is restored.

If driving the route is not successful, OTDRs are set up at different sites, usually at the repeater huts, to find the breaks. Other alarms on the electronics can show if there are other types of degradation in the system. The operator at remote sites can measure a dc voltage level, which is related to the receiver's optical input power, to monitor if the received signal is normal.

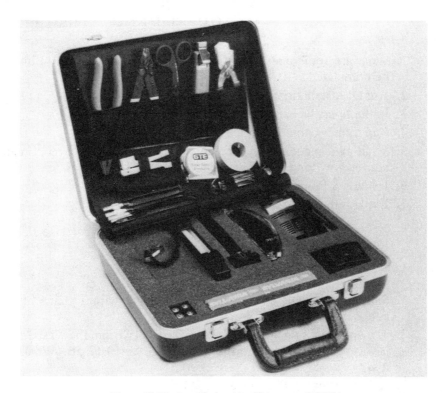

Figure 12-15   Installation kit. (Courtesy of GTE.)

12.14 INSTALLATION AND CONNECTORIZATION KITS

Installation and connectorization kits are used routinely by the field installer. The kits contain all the tools commonly needed to open and prepare fibers, to install and polish connectors, and to inspect them. Some of the tools are specific to the fiber being used; others are as common as a screwdriver. The installation kit is shown in Figure 12-15.

12.15 SUMMARY

All the installation methods discussed in this chapter require that the fiber optic cable meet certain specifications for physical stress and environmental protection. Cable is designed with characteristics that make it suitable for a particular type of installation or application. The two most important pieces of test equipment are the power meter and an OTDR. These two pieces of equipment can help test the cable plant, the transmitter, and receiver for any fiber optic system.

QUESTIONS

1. What are some questions that need to be asked before an installation can be completed?
2. Are all installations the same? (Give an example.)
3. Is the fiber the same for all installations?
4. Can a fiber cable be pulled by the connectors?
5. What two pieces of equipment must be taken on every installation no matter how long it is?
6. Explain a loopback test.
7. Explain the cutback method.

PROBLEMS

1. Sketch an OTDR display for a 20 km link with a splice at 6 km and a connector at 2 and 15 km.
2. The splice loss on an OTDR was measured. From the west end of the fiber it was 0.25 dB and from the east end it was 0.12 dB. What is the splice loss?
3. A 5 m piece of fiber measured 3 dB power out. A 30 m piece of fiber measured 1.2 dB power out. What is the power loss due to connectors, splices, and imperfections in the fiber?
4. The measured power for a short length of fiber was -10 dB and for the long length of fiber -20 dB. What is the fiber loss, and what is the attenuation (dB/km) for a 20 km fiber cable?

Glossary

Absorption State in which a photon has been consumed by an electron in the valence band. (Chapter 6)

Acceptance Angle Range of angles within which an injected light beam will enter the fiber. (Chapter 3)

Acceptance Cone Acceptance angle in three dimensions. (Chapter 3)

Armoring One of the last layers of a fiber optic cable, added for protection against the rigors of installation, rodents, lightning; can act as a support of aerial cables. (Chapter 4)

Avalanche Multiplication Gain resulting from reverse biasing an avalanche photodiode. (Chapter 7)

Avalanche Photodiode Photodetector that uses a large reverse bias to create a gain or multiplication of the conversion of incident photons to electrons. (Chapter 7)

Bandgap (Energy Gap) Amount of energy needed for the electron to jump from the valence band (equilibrium state) to the conduction band (excited state). (Chapter 6)

Bandwidth Frequency ranges for a system. (Chapter 3)

Baud Rate Signaling speed. (Chapter 8)

Bending Losses Losses due to imperfections in the fiber. (Chapter 3)

Buffer Tube Used to enhance the tensile strength and provide radial protection for the fiber. (Chapter 4)

Bus Network Topology where the nodes tap-off from a central backbone. (Chapter 10)

Cable Core Filling Material added to the cable to make it watertight. (Chapter 4)

Cable Sheath Provides an enclosure for the strength member, filling, fiber cables, and central member. (Chapter 4)

Central Member Helps with the stranding (braiding the fibers around a central point) and prevents buckling and kinking. (Chapter 4)

Chromatic Dispersion Sum of material and waveguide dispersion; found only in single-mode fiber. (Chapter 3)

Cladding Outer portion of the fiber optic cable, that confines the light. (Chapters 1 and 3)

Coatings Applied to the fiber at the drawing tower and used to enhance the strength of the optical fiber. (Chapter 4)

Core Inner portion of a fiber optic cable that light travels through. (Chapters 1 and 3)

Coupler Device used to split optical power into other fiber(s). (Chapter 5)

Coupler, Star Splits optical power among many fibers. (Chapter 5)

Coupler, T Three-port device that splits optical power. (Chapter 5)

Critical Angle Angle above which rays do not propagate in a fiber. (Chapter 3)

Cutoff Frequency Lowest frequency that will produce a photocurrent. (Chapter 7)

Cutoff Wavelength Wavelength at which modal or material dispersion becomes excessive. (Chapters 3 and 7)

Dark Current Current that arises when no incident light is falling on the photodiode. (Chapter 7)

Dispersion Pulse broadening of the signal at the received end of the fiber. *See also* Chromatic and modal dispersion. (Chapter 3)

Dispersion-Shifted Fiber Fiber specially made to have a minimum dispersion at 1550 nm rather than at 1300 nm. (Chapter 3)

Dynamic Range Difference in the minimum and maximum acceptable power levels. (Chapter 9)

Electroluminescence Converting electrical energy to light, such as in a forward-biased *pn* junction of a diode, thereby generating photons. (Chapter 6)

Electromagnetic Spectrum Designation of frequencies and wavelengths. (Chapter 2)

Electromagnetic Wave Wave that propagates by the interchange of energy between the electrical and magnetic fields. (Chapter 2)

Energy Gap *See* Bandgap.

Frequency-Division Multiplexing Takes different signals, assigns different frequencies to each, and sends them on a single channel. (Chapter 8)

Fresnel Reflections Reflections from the boundary of two different refractive indices, where refraction is dominant. (Chapter 2)

Injection Laser Diode Type of laser using a forward-biased semiconductor device emitting coherent radiation. (Chapter 6)

Law of Reflection Angle of incidence equals angle of reflection. (Chapter 2)

Law of Refraction Angle of incidence does not equal angle of reflection. (Chapter 2)

Leaky Modes Light loss in the cladding. (Chapter 3)

Light-Emitting Diode Semiconductor *pn* junction that emits incoherent light. (Chapter 6)

Local Area Network Communication system that links computers and other systems together in close proximity to each other, usually less than 1 kilometer. (Chapter 10)

Losses, Extrinsic Losses due to alignment problems with the joining fibers. (Chapter 5)

Losses, Intrinsic Losses due to internal manufacturing problems with the fiber. (Chapter 5)

Macrobending Cracks that appear the length of the fiber, usually caused in installation. (Chapter 3)

Metropolitan Area Network Links computers and other communication signals together in a city.

Microbending Cracks that appear when the cable exceeds the minimum bend radius. (Chapter 3)

Mode Field Diameter Describes the light entering and propagating in a fiber. (Chapter 3)

Modes Propagation of light energy allowable in the core of the fiber. (Chapter 3)

Modulation Modifying the signal in some way. (Chapter 8)

Monochromatic Single wavelength or color. (Chapter 6)

Multiplexing Sending multiple signals over one channel. (Chapter 1 and 8)

Noise Equivalent Power Amount of input power needed to generate a photo-current equal to the noise current. (Chapter 7)

Normalized Wave Number (*V* number) Function of the fiber radius, NA, and wavelength of operation, used to calculate number of modes. (Chapter 3)

Numerical Aperture Light-gathering ability of a fiber. (Chapters 2 and 3)

Nyquist Rate Signal value sampled at twice the frequency. (Chapters 1 and 8)

Optical Time-Domain Reflectometer (OTDR) Piece of test equipment used to measure fiber attenuation, splice, and connector loss by using the backscatter of light. (Chapter 12)

Outer Jacket Used to protect the fiber and inner layers of the cable from moisture. (Chapter 4)

Photodetector Device that converts optical energy to electrical energy. (Chapter 7)

Photon Packet of light energy. (Chapter 2)

Pin Photodiode structure with three layers: a positive region, an intrinsic region, and a negative region. (Chapter 7)

Plenum Space between the ceiling, walls, or raised floor. (Chapter 12)

Preform Glass rod composed of core and cladding material which is drawn out to become a fiber. (Chapter 4)

Pulse Amplitude Modulation Changes the signal sent by its amplitude. (Chapter 8)

Pulse Code Modulation Modulation scheme used for telephony; signal is sampled and digitized. (Chapter 8)

Pulse Position Modulation Changes the signal sent by the level it is at at the position it is at. (Chapter 8)

Pulse Width Modulation Changes the signal sent by the width of the bit period. (Chapter 8)

Quantum Efficiency Efficiency of the conversion of incoming photons to electron–hole pairs. (Chapter 7)

Receiver, Optical Converts optical signals to a usable electrical output. (Chapter 9)

Reflection Light beam incident on a surface returns at the same angle. (Chapter 2)

Refraction Light beam enters a material from another; light is bent in the second material. (Chapter 2)

Refractive Index Ratio of the speed of light in a vacuum to the speed of light in a material. (Chapter 3)

Refractive Index Profile Manner in which the refractive index varies in the fiber construction from the center of the fiber to the outer part of the cladding. (Chapter 3)

Repeater Receiver and transmitter in series; The receiver takes the optical signal, regenerates it to a higher level electrically, and then transmits the signal again. (Chapter 9)

Resonant Cavity Confined region in a semiconductor device where stimulated emission occurs. (Chapter 6)

Responsivity Parameter that compares the output photocurrent to the input optical power. (Chapter 7)

Ring Network Topology where all of the nodes are linked together in a circle or ring fashion. (Chapter 10)

Scattering Light reflected in all directions due to imperfections in the fiber. (Chapter 3)

Shot Noise Random noise generated by photons generating optical current. (Chapter 7)

Signal-to-Noise Ratio Ratio comparing the amount of signal power to the noise power. (Chapter 7)

Spectral Width Narrow region of wavelengths that an ILD or LED emits. (Chapter 6)

Splice Permanent joining of two fibers. (Chapter 5)

Splice, Elastomeric Type of mechanical splice. (Chapter 5)

Splice, Fusion Method by which two fibers are joined together using extremely high heat. (Chapter 5)

Splice, Mechanical Two fiber ends are held together in a shell. (Chapter 5)

Spontaneous Emission Excited electron in the conduction band returns to the valence band and releases energy in the form of a photon. (Chapter 6)

Star Network Topology scheme where the nodes radiate out from a central hub, usually in a starlike fashion. (Chapter 10)

Stimulated Emission External photon excites an electron in the conduction band to fall to the valence band, thus releasing a photon with the same wavelength as the external photon. (Chapter 6)

Stranding Method by which the fibers are wound around the central member. (Chapter 4)

Strength Member Used to add tensile strength to a cable for installation purposes. (Chapter 4)

T-Carrier Multiplexing rates used by the phone company. (Chapter 1)

Thermal Noise Random noise generated by the temperature of the photodetector. (Chapter 7)

Time-Division Multiplexing Assigns a certain time frame in which a signal can be put on the single channel. (Chapter 8)

Topology Types of networking schemes laid out in a ring, star, bus, or tree scheme. (Chapter 10)

Total Internal Reflection Ray of light bouncing at angles in a confined tube; principle on which fiber optics is based. (Chapter 2)

Transceiver Sends and receives a signal, usually over two separate lines. (Chapter 9)

Transmission Windows Region of wavelengths that have low loss for the fiber and emitting devices. (Chapter 6)

Transmitter, Optical Converts electrical signals to optical signals. (Chapter 9)

V **Number** *See* Normalized wave number. (Chapter 3)

Wavelength-Division Multiplexing Multiplexing scheme that takes different signals, sends them with different wavelengths, and passes them through a single channel. (Chapter 8)

Wide Area Network Links communication networks together in a large area, usually between cities and sometimes even states.

Appendix A:

Constants and Decibels

CONSTANTS

The values of constants used in this book are as follows:

$h = 6.626 \times 10^{-34}$ Joule second   Planck's constant

$e = 1.602 \times 10^{-19}$ coulombs   electron charge

$c = 2.998 \times 10^{8}$ m/s   speed of light in a vacuum

$K = 1.381 \times 10^{-23}$ Ws/K   Boltzmann's constant

DECIBELS

Decibels are used as a power measurement. Ordinary power gain is defined as the output power divided by the input power. Decibel gain is defined as 10 times the logarithm of the ordinary power gain. The decibel is used because each time the ordinary gain increases by a power of 10, the decibel gain increases by 10. Putting it simply, by using decibel gain, the student only has to add 10 each time the power is increased by a factor of 10.

dB Values

There is a lot of confusion about optical versus electrical power decibel values. The following conversions apply. Optical power P_{opt} is converted to electrical current I in a detector.

$$\text{dB (optical)} = 10 \log \left(\frac{P_{\text{opt1}}}{P_{\text{opt2}}} \right)$$

$$= 10 \log \left(\frac{I_1}{I_2} \right)$$

$$\text{dB (electrical)} = 10 \log \left(\frac{P_{e1}}{P_{e2}} \right)$$

$$= 20 \log \left(\frac{I_1}{I_2} \right)$$

Therefore,

$$\text{dB (electrical)} = 2 \times \text{dB (optical)}$$

dBm Values

There is also confusion over using dB versus dBm in fiber optics. dBm is used mainly because the received power is so small that the dB value would also be small. One milliwatt of optical power is used as a reference when specifying absolute optical power level P_{opt} in terms of dBm.

$$\text{dBm (optical)} = 10 \log \left(\frac{P_{\text{opt}}}{1 \text{ mW}} \right)$$

These values are important in establishing the operating levels of the transmitter, receiver, cables, connectors, and splices.

Power in milliwatts can be derived from the following formula:

$$P(\text{mW}) = 10^{\text{dBm}/10}$$

Appendix B:
Standards

This appendix considers only a few of the major contributors to the standardization process.

AMERICAN NATIONAL STANDARDS INSTITUTE (ANSI)

FDDI	Fiber Distributed Data Interface
T1.101-1987	Synchronous Interface Standards for Digital Networks
T1.102-1987	Digital Hierarchy Electrical Interfaces
T1.103-1987	
T1.106-1988	Digital Hierarchy SYNC DS-3 Format Spec
T1.107-1988	Digital Hierarchy Optical Interface
draft T1.105-1988	Specs: Single-Mode
	Digital Hierarchy Format Specs
	Synchronous Optical Network (SONET)

BELLCORE PUBLICATIONS

TA-TSY-000442	Generic Criteria for Fiber Optic Couplers
TA-NPL-000449	Generic Requirements and Design Considerations for Fiber Distributed Frames
TA-TSY-000917	SONET regenerator

CCITT DOCUMENTS

Recomm. M.30	Principles for a Telecommunication Management Network
Recomm. G.703	Physical/Electrical Characteristics of Hierarchical Digital Interfaces
Recomm. Q.921 (I.441)	ISDN User-Network Interface Data Link Layer Specification

DEPARTMENT OF DEFENSE

DOD-STD-1678	Fiber Optic Test Methods
DOD-C-85045	Fiber Optic Cable
FSC 6625	Test Equipment
MIL-HDBK-141	Fiber Optic Design Practices

ELECTRONICS INDUSTRIES ASSOCIATION (EIA)

Following is a list of the subcommittees within the EIA.

FO-2 Optical Communications Systems

FO-2.1	Optical Fiber Telecom Systems
FO-2.2	Fiber Optics LAN
FO-2.3	Jitter and Wander
FO-2.5	Cable Plant Installation

FO-6 Fiber Optics

FO-6.1	Field Tooling and Test
FO-6.2	Terminology Definition and Symbology
FO-6.3	Interconnection Devices
FO-6.4	Test Methods and Instrumentation
FO-6.5	Fiber Optic Transducers
FO-6.6	Optical Fibers and Materials
FO-6.7	Fiber Optic Cables

RS-455-XXX Fiber-Optic Test Procedures

The FOTP-XXX test procedures number from 1 to 179 and include every type of testing from connectors to fiber to terminal equipment. A list is not included here but can be obtained from the EIA.

Other EIA Documents

RS-440-1978	Connector Terminology
RS-458,-459	Standards for Fiber Classes and Materials
472A-XX0-86	Cables for Outside Aerial
472B-XX0-86	Cables for Underground and Burial
472C-XX0-86	Cables for Indoor Use
472D-XX0-86	Cables for Outside Plant Use
RS-475-1986	Generic Spec. Connectors
RS-492-1987	Generic Spec. Optical Waveguide
RS-509-1984	Generic Spec. Fiber Optic Terminal Equipment
RS-515-1986	Generic Spec. Fiber and Cable Splices
TSB-20	Single Mode Fiber Optic System Transmission Design

INSTITUTE OF ELECTRICAL AND ELECTRONIC ENGINEERS (IEEE)

ANSI/IEEE STD 100-1984	IEEE Standard Dictionary of Electrical and Electronic Terms
IEEE 812-1984	Definition of Terms Relating to Fiber Optics

NATIONAL BUREAU OF STANDARDS (NBS)

Special Pub. 637	Optical Fiber Characterization

U.S. GOVERNMENT PUBLICATIONS

21 CFR 1040	Performance Standards for Laser Products

Bibliography

CHAPTER 1: INTRODUCTION TO FIBER OPTICS

BARNOSKI, MICHAEL. "1990 Trends in Fiber Optics: Coming Advances in Photonics." *Photonics Spectra*. January 1990.

CCITT. *Optical Fibres for Telecommunications*. The International Telegraph and Telephone Consultive Committee, Geneva. 1984.

CHERIN, ALLEN. *An Introduction to Optical Fibers*. McGraw-Hill, New York. 1983.

HECHT, JEFF. *Understanding Fiber Optics*. Howard W. Sams, Indianapolis, Ind. 1989.

METCALF, BRUCE D. AND CHARLES W. KLEEKAMP. *Fiber Optic Communications*. Course Notes. September 1982.

MORRIS, DAVID J. *Pulse Code Formats for Fiber Optical Data Communication, Basic Principles and Applications*. Marcel Dekker, New York. 1983.

PALAIS, JOSEPH C. *Fiber Optic Communications*. Prentice Hall, Englewood Cliffs, N.J. 1984.

STREMLER, FERREL G. *Introduction to Communication Systems*. Addison-Wesley, Reading, Mass. 1977.

Technician's Guide to Fiber Optics. AMP, Inc. and Delmar Publishers, Albany, N.Y. 1987.

ZANGER, HENRY AND CYNTHIA ZANGER. *Fiber Optics Communication and Other Applications*. Merrill Publishing Co., Westerville, Ohio. 1991.

CHAPTER 2: THE PRINCIPLES OF LIGHT

HECHT, JEFF. *Understanding Fiber Optics.* Howard W. Sams, Indianapolis, Ind. 1989.

JENKINS, FRANCIS A. AND HARVEY E. WHITE. *Fundamentals of Optics,* Fourth Edition. McGraw-Hill, New York. 1976.

LACY, EDWARD A. *Fiber Optics.* Prentice Hall, Englewood Cliffs, N.J. 1982.

MAHLKE, GUNTHER AND PETER GOSSING. *Fiber Optic: Cable Fundamentals, Cable Technology, and Installation Practice.* Wiley, Chichester, West Sussex, England. 1987.

METCALF, BRUCE D. AND CHARLES W. KLEEKAMP. *Fiber Optic Communications.* Course Notes. September 1982.

STERLING, DONALD J., JR. *Technician's Guide to Fiber Optics.* Delmar Publishers, Albany, N.Y. 1987.

Technician's Guide to Fiber Optics. AMP, Inc. and Delmar Publishers, Albany, N.Y. 1987.

YOUNG, HUGH D. *Fundamentals of Waves, Optics, and Modern Physics.* McGraw-Hill, New York. 1976.

ZANGER, HENRY AND CYNTHIA ZANGER. *Fiber Optics Communication and Other Applications.* Merrill Publishing Co., Westerville, Ohio. 1991.

CHAPTER 3: OPTICAL FIBER AND ITS PROPERTIES

CAMPBELL, LARRY D. *Fiber Optic Communication Systems Course Notes.* Integrated Computer Systems. September 1980.

Corguide Optical Fiber Measurement Method. "Mode-Field Diameter." Issued 10/87.

ESTY, SCOTT A. "Transmitting Video by Light." *Installer Technician.* December 1988.

GOWAR, JOHN. *Optical Communication Systems.* Prentice Hall International, Hemel Hempstead, Hertfordshire, England. 1985.

HENTSCHEL, CHRISTIAN. *Fiber Optics Handbook.* Hewlett-Packard, Germany. October 1983.

MAHLKE, GUNTHER AND PETER GOSSING. *Fiber Optic: Cable Fundamentals, Cable Technology, and Installation Practice.* Wiley, Chichester, West Sussex, England. 1987.

METCALF, BRUCE D. AND CHARLES W. KLEEKAMP. *Fiber Optic Communications.* Course Notes. September 1982.

PALAIS, JOSEPH C. *Fiber Optic Communications.* Prentice Hall, Englewood Cliffs, N.J. 1984.

RODDY, DENNIS AND JOHN COOLEN. *Electronic Communications,* Third Edition. Reston Publishing, Reston, Va. 1984.

SENIOR, JOHN M. *Optical Fiber Communications, Principles and Practices.* Prentice Hall International, Hemel Hempstead, Hertfordshire, England. 1985.

STERLING, DONALD J., JR. *Technician's Guide to Fiber Optics.* Delmar Publishers, Albany, N.Y. 1987.

STOROZUM, STEPHEN L. "Fiber Optic Systems: Practical Design III." *Photonics Spectra.* November 1985.

SZANTO, J. ATTILA. "Fiber Optic Basics (Components, Waveguides and Some Cable Construction)." *Fiber Optic Applications in Electrical Power Systems.* IEEE Tutorial Course. 1985. pp. A1–A6.

Technician's Guide to Fiber Optics. AMP, Inc. and Delmar Publishers, Albany, N.Y. 1987.

CHAPTER 4: FIBER FABRICATION AND CABLE DESIGN

BAKER, DONALD G. *Fiber Optic Design and Applications.* Reston Publishing, Reston, Va. 1985.

BASCH, E. E. BERT, editor. *Optical Fiber Transmission.* Howard W. Sams, Indianapolis, Ind. 1987.

CAMPBELL, LARRY D. *Fiber Optic Communication Systems.* Course Notes: Integrated Computer Systems. September 1980.

CLARK, ROBERT S., editor. *The Photonics Design and Application Handbook 1985.* Optical Publishing Company, Pittsfield, Mass. 1985.

Designers Guide to Fiber Optics. AMP, Inc., Harrisburg, Pa. 1982.

ELION, GLENN R. AND HERBERT ELION. *Fiber Optics in Communication Systems.* Marcel Dekker, New York. 1978.

JONES, RICHARD J. *Optical Fiber Communication Systems.* Course Notes, Volume I. Siecor FiberLAN and North Carolina State University, Raleigh, N.C. 1984.

KAO, CHARLES K. *Optical Fiber Systems: Technology, Design, and Applications.* McGraw-Hill, New York. 1982. pp. 63–68.

MAHLKE, GUNTHER AND PETER GOSSING. *Fiber Optic: Cable Fundamentals, Cable Technology, and Installation Practice.* Wiley, Chichester, West Sussex, England. 1987.

MARROW, A. J. "Advances in Vapor-Phase Processing Techniques." *Fiberoptic Product News.* January 1988.

MILLER, STEWART E. AND ALAN G. CHYNOWETH. *Optical Fiber Telecommunications.* Academic Press, New York. 1979.

NEROU, JEAN PIERRE. *Introduction to Fiber Optics.* Griffon D'Argile, Sainte-Foy, Quebec 1988.

PALAIS, JOSEPH C. *Fiber Optic Communications*. Prentice Hall, Englewood Cliffs, N.J. 1984.

SANDBANK, C. P., editor. *Optical Fibre Communication Systems*. Wiley, New York. 1980.

SENIOR, JOHN M. *Optical Fiber Communications, Principles and Practices*. Prentice Hall International, Hemel Hempstead, Hertfordshire, England. 1985.

SZANTO, J. ATTILA. "Fiber Optic Basics (Components, Waveguides and Some Cable Construction)." *Fiber Optic Applications in Electrical Power Systems*. IEEE Tutorial Course. 1985. pp. A1–A6.

CHAPTER 5: CONNECTORS, SPLICES, AND COUPLERS

BOWES, KENNETH. "Fiber Connectors: A Primer for the Wary Designer." *Photonics Spectra*. October 1989.

CAMPBELL, LARRY D. *Fiber Optic Communication Systems Course Notes*. Integrated Computer Systems. September 1980.

CLARK, ROBERT S., editor. *The Photonics Design and Application Handbook 1985*. Optical Publishing Company, Pittsfield, Mass. 1985.

Designers Guide to Fiber Optics. AMP, Inc., Harrisburg, Pa. 1982.

DREXEL, R. PATRICK, JAMES A. NELSON, AND JOHN SCHNECKER. "New Approaches to Termination." *Photonics Spectra*. March 1988.

ELION, GLENN R. AND HERBERT ELION. *Fiber Optics in Communication Systems*. Marcel Dekker, New York. 1978.

GILLIAM, FREDERICK AND HAROLD A. ROBERTS. "Designer's Handbook: Fiber Optic Couplers for Multiplexing." *Photonics Spectra*. April 1984.

HECHT, JEFF. "Fiber Splicing Made Easy." *Fiber Optics Product News*. May 1987.

———— *Understanding Fiber Optics*. Howard W. Sams, Indianapolis, Ind. 1989.

JALBERT, KARL. "Recognizing Connector Variations and Their Fiber Optic Applications." *Fiber Optics Product News*. September 1987.

JONES, RICHARD J. *Optical Fiber Communication Systems*. Course Notes, Volume I. Siecor FiberLAN and North Carolina State University, Raleigh, N.C. 1984.

KNECHT, DENNIS M. ET AL. Fiber Optic Field Splice. Internal article, Siecor Corporation, Hickory, N.C. December 14, 1981.

KOTELLY, GEORGE. "Special Report: Fiber Optic Connectors—Steady Progress and Performance." *Lightwave*. April 1989.

MAHLKE, GUNTHER AND PETER GOSSING. *Fiber Optic: Cable Fundamentals, Cable Technology, and Installation Practice*. Wiley, Chichester, West Sussex, England. 1987.

"Optical Communications." *Telecom Report,* Special Issue. Volume 6. Siemans. 1983.

SANDBANK, C. P., editor. *Optical Fibre Communication Systems.* Wiley, New York. 1980.

SEIPPEL, ROBERT G. *Fiber Optics.* Reston Publishing, Reston, Va. 1984.

SZANTO, J. ATTILA. "Fiber Optic Basics (Components, Waveguides and Some Cable Construction)." *Fiber Optic Applications in Electrical Power Systems.* IEEE Tutorial Course. 1985. pp. A1–A6.

Technician's Guide to Fiber Optics. AMP, Inc. and Delmar Publishers, Albany, N.Y. 1987.

CHAPTER 6: OPTICAL SOURCES

BRILEY, BRUCE E. *An Introduction to Fiber Optics System Design.* Elsevier, New York, 1988.

Designers Guide to Fiber Optics. AMP, Inc., Harrisburg, Pa. 1982.

HENTSCHEL, CHRISTIAN. *Fiber Optics Handbook.* Hewlett-Packard, Germany. October 1983.

Introduction to Fiber Optics and AMP Fiber-Optic Products. AMP, Inc., Harrisburg, Pa. 1979.

JONES, RICHARD J. *Optical Fiber Communication Systems.* Course Notes, Volume I. Siecor FiberLAN and North Carolina State University, Raleigh, N.C. 1984.

NEROU, JEAN PIERRE. *Introduction to Fiber Optics.* Griffon D'Argile, Sainte-Foy, Quebec 1988.

OLSEN, GREGORY H. "A Challenge to Laser Diodes, Long-Wavelength LED's." *Photonics Spectra.* September 1985. pp. 121–124.

STERLING, DONALD J., JR. *Technician's Guide to Fiber Optics.* Delmar Publishers, Albany, N.Y. 1987.

Technician's Guide to Fiber Optics. AMP, Inc. and Delmar Publishers, Albany, N.Y. 1987.

CHAPTER 7: PHOTODETECTORS

BRILEY, BRUCE E. *An Introduction to Fiber Optics System Design.* Elsevier, New York, 1988.

Designers Guide to Fiber Optics. AMP, Inc., Harrisburg, Pa. 1982.

FARRE, THOMAS R., editor. "1986–1987 Buying Guide, Technical Handbook and Industry Directory." *Fiberoptic Product News.* Volume 1, Number 5. 1986.

JONES, RICHARD J. *Optical Fiber Communication Systems*. Course Notes, Volume I. Siecor FiberLAN and North Carolina State University, Raleigh, N.C. 1984.

KAO, CHARLES K. *Optical Fiber Systems: Technology, Design, and Applications*. McGraw-Hill, New York. 1982. pp. 63–68.

METCALF, BRUCE D. AND CHARLES W. KLEEKAMP. *Fiber Optic Communications*. Course Notes. September 1982.

NEROU, JEAN PIERRE. *Introduction to Fiber Optics*. Griffon D'Argile, Harrisburg, Pa. 1988.

STERLING, DONALD J., JR. *Technician's Guide to Fiber Optics*. Delmar Publishers, Albany, N.Y. 1987.

SZANTO, J. ATTILA. "Fiber Optic Basics (Components, Waveguides and Some Cable Construction)." *Fiber Optic Applications in Electrical Power Systems*. IEEE Tutorial Course. 1985. pp. A1–A6.

SZE, S. M. *Physics of Semiconductor Devices*. Wiley, New York. 1969.

WYMER, SUSAN LEE. "Study of the Avalanche Multiplication and Signal-to-Noise Power Ratio in the Ternary $In_xGa_{1-x}As$ Avalanche Photodiode." Master's thesis, University of Central Florida. 1979.

CHAPTER 8: MODULATION SCHEMES FOR FIBER OPTICS

MORRIS, DAVID J. "Code Your Fiber-Optics Data for Speed without Losing Circuit Simplicity." *Electronic Design*. October 25, 1978.

———— *Pulse Code Formats for Fiber Optical Data Communication, Basic Principles and Applications*. Marcel Dekker, New York. 1983.

NEROU, JEAN PIERRE. *Introduction to Fiber Optics*. Griffon D'Argile, Sainte-Foy, Quebec 1988.

STREMLER, FERREL G. *Introduction to Communication Systems*. Addison-Wesley, Reading, Mass. 1977.

TAUB, HERBERT AND DONALD L. SCHILLING. *Principles of Communication Systems*. McGraw-Hill, New York. 1971.

CHAPTER 9: PRACTICAL OPTICAL TRANSMITTERS AND RECEIVERS

Designers Guide to Fiber Optics. AMP, Inc., Harrisburg, Pa. 1982.

MUOI, TRAN V. "Detectors and Receivers Reach for Sensitivity and Bandwidth." *Laser Focus World*. August 1988.

NEROU, JEAN PIERRE. *Introduction to Fiber Optics*. Griffon D'Argile, Sainte-Foy, Quebec 1988.

STERLING, DONALD J., JR. *Technician's Guide to Fiber Optics*. Delmar Publishers, Albany, N.Y. 1987.

Technician's Guide to Fiber Optics. AMP, Inc. and Delmar Publishers, Albany, N.Y. 1987.

CHAPTER 10: SYSTEMS ARCHITECTURE

BELLCORE. *Synchronous Optical Network (SONET) Transport Systems: Common Generic Criteria*. Technical Advisory TA-NWT-000253. Issue 6. September 1990.

GRANT, WILLIAM O. *Understanding Lightwave Transmission: Applications of Fiber Optics*. Harcourt Brace Jovanovich, San Diego, Calif. 1988.

HENRY, PAUL S. "High-Capacity Lightwave Local Area Networks." *IEEE Communications Magazine*. October 1989.

HOSS, ROBERT J. *Fiber Optic Communications Design Handbook*. Prentice Hall, Englewood Cliffs, N.J. 1990.

LIN, YIH-KANG MAURICE, DAN R. SPEARS, AND MIH YIN. "Fiber-Based Local Access Network Architectures." *IEEE Communications Magazine*. October 1989.

SENIOR, JOHN M. *Optical Fiber Communications, Principles and Practices*. Prentice Hall International, Hemel Hempstead, Hertfordshire, England. 1985.

SHERMAN, KEN. *Data Communications A User's Guide*. Prentice Hall, Englewood Cliffs, N.J. 1990.

Technician's Guide to Fiber Optics. AMP, Inc. and Delmar Publishers, Albany, N.Y. 1987.

CHAPTER 11: SYSTEM DESIGN

BAKER, DONALD G. *Fiber Optic Design and Applications*. Reston Publishing, Reston, Va. 1985.

BASCH, E. E. ET AL. "Calculate Performance into Fiber Optic Links." *Electronic Design*. August 16, 1980.

Basic Experimental Fiber Optic Systems. Motorola Technical Note T-52-01. Motorola Semiconductor Division. 1978.

BENDER, ALBERT AND STEVEN STOROZUM. "Charts Simplify Fiber-Optic Design." *Electronics*. November 23, 1978.

GRANT, WILLIAM O. *Understanding Lightwave Transmission: Applications of Fiber Optics*. Harcourt Brace Jovanovich, San Diego, Calif. 1988.

HENTSCHEL, CHRISTIAN. *Fiber Optics Handbook*. Hewlett-Packard, Germany. October 1983.

Hoss, Robert J. *Fiber Optic Communications Design Handbook*. Prentice Hall, Englewood Cliffs, N.J. 1990.

ITT Technical Note R-1. ITT Electro-Optical Products Division. Roanoke, Va. 1977.

Jones, Richard J. *Optical Fiber Communication Systems*. Course Notes, Volume I. Siecor FiberLAN and North Carolina State University, Raleigh, N.C. 1984.

Senior, John M. *Optical Fiber Communications, Principles and Practices*. Prentice Hall International, Hemel Hempstead, Hertfordshire, England. 1985.

Storozum, Stephen L. "Fiber Optic Systems: Practical Design I, II, III." *Photonics Spectra*. September, October, November 1985.

Technician's Guide to Fiber Optics. AMP, Inc. and Delmar Publishers, Albany, N.Y. 1987.

Transmission Systems for Communications, Fifth Edition. Bell Laboratories, Murray Hill, N.J. 1982.

CHAPTER 12: INSTALLATION AND TESTING OF FIBER SYSTEMS

Baker, Donald G. *Fiber Optic Design and Applications*. Reston Publishing, Reston, Va. 1985.

Bark, Peter R. et al. "Design, Testing, and Installation Experiences of a 35 Fiber Minibundle Submarine Cable." Presented at the *IEEE International Conference on Communications*, Philadelphia. June 1982.

Cunningham, D. J. *Cable Selection Questionaire*. Siecor Corporation, Research Triangle Park, N.C. March 1984.

Fairaizi, Alan F. *How to Select Fiber Optic Cables for Practical Applications*. Siecor Optical Cables, Inc., Horseheads, N.Y.

Fairaizi, Alan F. and Roosevelt A. Fernandes. *A Major Fiber Optic Test Bed for the Electric Utility Industry*. Siecor Corporation, Hickory, N.C. May 1980.

Fichtl, D. A. "A Checklist of Tips for Your Next Fiber Optic Pull." *Outside Plant*. May 1986.

Gentile, John. "Characterizing Optical Fibers with an OTDR." *Electro-Optical Systems Design*. April 1981.

Grant, William O. *Understanding Lightwave Transmission: Applications of Fiber Optics*. Harcourt Brace Jovanovich, San Diego, Calif. 1988.

Guide to Fiber Optic Installation. Belden Electronic Wire and Cable, Cooper Industries, Inc., Richmond, In. 1986.

Hentschel, Christian. *Fiber Optics Handbook*. Hewlett Packard, Germany. October 1983.

"Optical Communications." *Telecom Report,* Special Issue. Volume 6. Siemans. 1983.

"Optical Time Domain Reflectometry—Pinpointing Fiber Faults." *Test and Measurement World.* January 1983.

QUINN, G. C., editor. "Optical Time Domain Reflectometers." *Electronics Test.* June 1982.

Technician's Guide to Fiber Optics. AMP, Inc. and Delmar Publishers, Albany, N.Y. 1987.

APPENDIX A

Fiber Optic Technical Bulletin. Belden Corporation, Fiber Optic Group. Richmond, In. 1986.

HENTSCHEL, CHRISTIAN. *Fiber Optics Handbook.* Hewlett-Packard, Germany. October 1983.

Index

LED (*cont.*)
  power output, 109–10
  spectral width, 110–11
Light, 16
Liquid-phase deposition, 60
Local area network (LAN), 163–64
Loss budget, 181–83
Losses, 77–80
  excess, 94
  extrinsic, 79–80
  insertion, 93
  intrinsic, 77–79
  splitting, 93
Low-impedance receiver, 156–57

M

Macrobending, 48
MAN, 165–66
Manchester coding, 137
Material, 29–31
  glass, 29
  plastic, 31
  plastic clad silica, 29–31
Maurer, Robert, 3
Medical industry, 9
Metropolitan area network (MAN), 165–66
Microbending, 47
Military, 8
Mode field diameter, 49
Modes, 32–34
  leaky, 47
Mode stripping, 47
Modified chemical vapor deposition, 61–62
Modulation schemes, 135–46
Monomode fiber, 37
Multimode graded-index fiber, 38
Multimode step-index fiber, 36
Multiple access schemes, 163–65
  frequency division multiple access, 164
  time division multiple access, 164
Multiplexing, 135

N

NEP, 128–29
Networks, 162–68
  local area network (LAN), 163–64
  metropolitan area network (MAN), 165–66
  point-to-point, 162–63
  testing, 208–10
  wide area network (WAN), 166–68
Noise, 124–25
  quantum, 124
  shot, 124
  thermal, 125
  total, 125
Noise equivalent power, 128–29
Non-return to zero, 136
Normalized frequency cutoff, 33
Normalized wave number, 33
Nuclear radiation, 6
Numerical aperture, 22–23
Numerical aperture mismatch, 78

O

Optical receiver, 154–58
  amplifier, 155, 158
  low-impedance, 156–57
Optical time domain reflectometer, (OTDR), 206–8
Optical transmitters, 147–54
  analog ILD, 152–54
  analog LED, 152
  digital ILD, 152–54
  digital LED, 148–49
Oscilloscopes, 203–4
OTDR, 206–8
Outside vapor deposition, 62–63

P

Packaging, 159
Patch panel, 199

Photodetectors, 118–34
  avalanche photodiodes, 120–21
  PIN, 119–20
Photon, 15–16, 100
Photophone, 2
Planck's constant, 15
Plasma-activated vapor deposition, 63
Plasma-impulse vapor deposition, 63
Plastic-clad silica, 29–30
Plastic fiber, 50
Point-to-point, 162–63
Polishing methods, 81
Population inversion, 101
Power budgets, 176–91
Power meters, 204–5
Power throughput analysis, 181–83
Preform, 58
Profile mismatch, 79
Pulse amplitude modulation, 139–40
Pulse code modulation (PCM), 141–42
Pulse position modulation, 140
Pulse width modulation, 140

Q

Quantum efficiency, 104, 122

R

Rayleigh scattering, 46, 206
Receiver sensitivity, 154–55
Reflections, 17
Refractive index, 16
Refractive index profile, 34
Reliability, 112
Repeater, 6, 158
Repeater spacing analysis, 185–87
Resonant cavity, 106
Responsivity, 123
Return-to-zero, 137
Ring topology, 163
Rise-time analysis, 183–85
Rod-in-tube, 59
Round trip delay, 190
RS-455, 201

S

Satellite communications, 9
Scattering, 46
  Rayleigh, 46
Schultz, Peter, 2
Self-healing LAN, 165
Signal-to-noise ratio, 126–27
Singlemode fiber, 37
Snell's Law, 19
SNR, 126–27
SONET, 172
Sources, 99–117
Spectral width, 110–11
Speed of response, 126
Splice closure, 199
Splices, 82–85
  Fastomeric, 85
  fusion, 84
  mechanical, 83–84
Split phase modulation, 137
Spontaneous emission, 101–2
Spot size, 49
Standards, 168–72
Star topology, 163
Stimulated emission, 101
Stranding, 69–70
  helical, 69
  reverse-lay, 70
Surveillance, 8
Switches, 95–96
  bypass, 96
  two position, 95
SYNTRAN, 171
System design, 181–85
  LAN loss budget, 187–89
  power throughput analysis, 181–83
  repeater spacing, 185–187
  risetime analysis, 183–85

T

T-1, 4
T-carrier, 4
TDMA, 164
Telephone industry, 7–8
Television industry, 8